Wild Animals of Western Canada

**An Altitude
NatureBook**

Wild Animals OF Western Canada

Kevin Van Tighem

Altitude Publishing
Canadian Rockies/Vancouver

Photo, front cover:
Grizzly Bear

Photo, back cover:
Killer Whale

Copyright © 1992
Altitude Publishing Ltd.
1500 Railway Avenue,
Canmore, Alberta
Canada T1W 1P6

**Canadian Catalogue in
Publication Data**
Van Tighem, Kevin J.
Wild Animals of Western Canada
Include bibliographical references.
ISBN 1-55153-800-8
1. Zoology - Canada, Western -
Guidebooks.
2. Animals - Canada, Western -
Guidebooks.
I. Title
QL221.W38V35 1992 591.9712
C92-091354-7

9 8 7 6 5

Editor: John F. Ricker
Design: Robert MacDonald, Media Clones Inc.
Cover design: Stephen Hutchings

Made in Western Canada
Printed and bound in Western Canada
by Friesen Printers, Altona, Manitoba

Altitude Publishing Canada Ltd.
gratefully acknowledges the support
of the Canada/Alberta Agreement
on cultural industries.

Photographic Credits:
Mark Hobson: 87, 93
Doug Leighton: 2, 11, 13, 19,
   27, 28, 33, 37, 42, 55, 57,
   67, 68-9, 80-1
Ricardo Ordóñez: back cover
Graeme Pole/Mountain Vision:
   43
Dennis Schmidt: front cover, 9,
   17, 25, 29, 44, 47, 48-9,
   52, 53, 61, 70-1, 73, 77,
   83, 89, 91
Esther Schmidt: 15, 21, 23, 31,
   34-5, 39, 40, 41, 51, 56,
   59, 63, 65, 75, 79, 85

Altitude GreenTree Program
Altitude will plant in Canada twice as many
trees as were used in the manufacturing of
this book.

Contents

Introduction

High on Canada's Continental Divide the vast Columbia Icefield spills its glaciers down lonely Rocky Mountain valleys, their meltwaters flowing at length to three different oceans. The short summers in this cold and stormy place barely give wildlife time to recover from one long winter and store fat for the next.

Mountain goats live amid the stunted trees and shrubs that grow on cliffs above the ice. Where cold winds draining from the glaciers sweep snow from nearby slopes, small bands of bighorn sheep dwell and range up into the high meadows. Down on the flats of the Sunwapta River, caribou forage in the spruce forests and grizzly bears leave tracks in the river mud.

It is a spectacular world, but a hostile one. Where glaciers occur, few animals reside. Yet, it was not that long ago when nearly all of western Canada lay frozen beneath glacier ice. Each summer, thousands of visitors flock to the Canadian Rockies to visit the remnants of these once great rivers of flowing ice. But only a few realize that in making the journey to the Columbia Icefield, they have been backtracking the very glaciers they came to see.

Virtually all of western Canada bears the marks of the glaciers that scoured and shaped the landscape until a few thousand years ago. The rugged west coast, the mountains, the northern forest, the parkland, and the prairies of western Canada are relatively new, as earth measures time.

Everything one sees here today has arrived only in the past few thousand years. In fact, the changes are still going on today; the land is still sorting itself out after millenia in a deep freeze.

On the west coast of British Columbia, the glaciers carved fiords and deep channels where whales and seals range today.

The great Fraser delta is the product of centuries of glacial erosion in the mountain ranges that form the headwaters of the Fraser River. Erosion continues to feed the mighty river with silt, sand, and gravel that have come to rest, over countless centuries, in the rich bottomlands of the delta.

Knife-edged ridges, broad valleys and steep slopes of glacial till carpeted with dense forests betray the work of glaciers in carving the mountain landscape. East of the Rockies the marks of ancient glaciers are more subtle. Forested eskers wind snake-like across the boreal forest, tracing the paths of glacial runoff streams, and still serving as paths for caribou and wolves. Deep potholes remain where great chunks of ice melted centuries ago; today, mink hunt muskrats around the edges of these muskeg ponds while moose forage for pondweeds in their depths.

Pronghorns race across the Saskatchewan prairie, oblivious to the fact that 15 thousand years ago huge sheets of ice smoothed out these plains and deposited the gravelly debris that lies beneath the prairie grassland. Coulees – great valleys that slice abruptly through the plains, often with little or no running water at the bottom – mark the channels of immense rivers that drained the melting ice 10 thousand years ago. Mule deer feed along the coulee slopes, while bobcats shelter in the shrubby tangles along the valley bottoms.

White-tailed deer grow fat on corn and alfalfa during southern Manitoba's long, hot summers. But the food they eat grows on silt that settled out of a vast glacial lake only a few thousand years ago. The same kind of silt blows down Jasper's Athabasca Valley each winter, building huge dunes near Jasper Lake where timber wolves dig their dens.

The highways and railroads that we late-comers use to explore the Canadian West criss-cross landscapes that still show the tracks of glaciers. Not only did the great glaciers shape the scenery, they helped to create the wildlife heritage for which western Canada is so famous. The West's diverse wildlife habitats – from sandhills to coulees, to fiords and floodplains and foothills – have evolved from the distinctive landforms and soils that are the legacy of glaciers.

Today, grizzly bears still roam the same high country that shelters the last great glaciers. The glaciers, however, are retreating in the face of a warming climate. The grizzlies, too, are retreating... in the face of a changing world.

In the past two centuries, an even more potent force for change than the glaciers has begun to rearrange the landscape again: western civilization.

The ghosts of western Canada's glaciers have a message for all of us: nothing is forever. The wildlife wealth of Canada's plains, mountains, and coastlands was not always here. It has evolved over 10 thousand years. Much of it has disappeared in the last hundred; there are very few wildlife populations in Canada that are not under threat today from global climate change, burgeoning human populations, and resource development projects on a scale unprecedented on this planet.

Even so, for North Americans, these are the best of times. From the comfort of your vehicle you can see 500 pronghorns race across the prairie, or watch a black bear and her cubs eat dandelions in the Rockies, or even spot a grey whale breaching off the west coast of Vancouver Island. We have freedom of movement, a vast and beautiful landscape to explore, and numerous wild animals to enlighten and inspire us. We also hold the future of this living mosaic in our hands.

Can we find the love and humility in our hearts to spend a little time learning to understand the wild things that share the West with us? It is up to each of us to prove that natural places, and the wildlife that depends on them can coexist with human beings who truly love, and thus truly belong in, the great, evolving ecosystems of western Canada.

Moose

JACK NOLAN AND I were hiking along a tributary to Jasper National Park's Fiddle River one July day when a reddish-brown animal jumped out of the willows and trotted off down the trail ahead of us.

"Moose calf," said Jack. "Look out!"

There was a loud crash, and an enormous black shape exploded from the timber just beyond where the calf had appeared. I breathed a sigh of relief as the huge cow moose wheeled and followed her baby down the trail and over a rise.

A moment later, however, she reappeared. Her eyes were bulging and long strings of saliva hung from her mouth. She made straight for us like an avenging angel.

I hadn't even known I could climb a tree, much less one with no branches; nonetheless, I was fifteen feet up a skinny spruce in less time than it takes to tell about it. The moose crashed back and forth in the undergrowth between my tree and the one that held Jack. Finally she turned and loped away after her calf.

I've never had such a close call with a wild animal – not even being treed by a grizzly or stared down by a big black bear. Nothing has left so vivid an impression on me as that encounter with an angry mother moose. A moose is a very big animal.

In fact, a moose is one of the largest land animals in North America, second only to the bison. A mature bull moose can weigh over 500 kilograms. Rare indeed is the moose hunter who, having shot one moose deep in the forest, will ever again hunt moose more than a few metres from a road or trail.

Moose live in mixed forests of conifers and deciduous trees, especially near areas that have been burned during the last few decades. They also prefer the wetland complexes that line northern rivers and lakeshores. Moose are browsers; they eat mostly leaves, buds, and twigs of deciduous shrubs, especially willows and poplars.

In the past few years, moose have become established in areas where they had not been seen for decades, or where they have never been seen before. The forested country along BC's Coquihalla Highway has a population of moose, even though this is not a traditional range. Similarly, moose have become established in the aspen parkland around Red Deer and Stettler and even in some of the irrigated farmland east of Calgary.

Moose have one calf each year. They usually retreat to an island in a river or to some other secluded spot where the cow can protect her calf from bears, wolves, and other predators. Wolves and cougars prey on moose most successfully in winter.

Modern logging practice opens up old forest stands with clearcuts and roads. Wilderness-dependent animals like caribou and, to a lesser degree, elk can suffer from this sort of industrial development. Responsible forestry can benefit moose through growing shrubby, young vegetation, while retaining large stands of old

... a moose is one of the largest land animals in North America, second only to the bison.

trees for shelter and protection. However, moose are relatively easy to hunt, so industrial roads can become a problem if they allow too many hunters access. Those interested in conserving moose populations recommend closing and reclaiming logging and other resource roads once they are no longer needed.

Elk

LATE IN AUGUST, as the grass fades to gold and the aspens start to change colour, you might hear a long, high-pitched squeal giving way to a series of deep grunts. The elk rut, or mating season, has begun. For the next month and a half the meadows and valleys will ring with the eerie bugling. In the shadows along the edges of aspen and fir forests, great bull elk with ivory-tipped antlers will pace stiff-legged around the edges of their herds of placid cows and calves, pausing now and then to extend their necks, curl back their lips, and fling another deafening challenge into the night.

Few events are as thrilling as to stalk a herd of elk in September and watch the herd bull display his massive antlers and bugle. On one occasion I was fortunate enough to watch as two bulls locked antlers and fought. This was no mere pushing match; clods of dirt flew, and antlers clattered as the antagonists dashed back and forth around the meadow. One elk finally fled with a gash in one shoulder where the other's antler tine had torn it. On another occasion a big bull chased me behind a tree. When he was unable to get around the tree faster than I could dodge, the bull contented himself with raking the foliage with his antlers. Wildlife photographers have been seriously injured by angry bulls. In the Jasper National Park information centre, two elk skulls suspended from the a stairwell ceiling offer mute proof that rutting elk sometimes battle to the death. In this case, both elk locked antlers and, unable to separate, eventually grew weak and died.

A friend who studied grizzlies in Banff's Cascade Valley made some of his best observations when a family of bears moved in to feed on a huge bull elk. The bull had snapped his neck when he tripped in pursuit of another bull during the rut.

By early October, with most of the cow elk bred, the herd bulls gradually drift away from their harems and begin to spend more time alone, or in peaceful coexistence with other bulls. By the time the first snows of winter carpet the landscape, the mature bulls almost never associate with the larger herds of cows, calves, and young bulls.

Bull elk go into the winter at a disadvantage; they generally have very little fat left after the exhausting ordeal of gathering, defending, and breeding their harems of cows. The cows, on the other hand, are fat and rested. Ecologically, this works out well. Once the cows have been bred,

All through the glory of the brief northern summer, elk grow sleek and fat on the season's green wealth ...

bulls are no longer important to the population. If, in their weakened state, they fall prey to wolves, a cougar, or starvation in deep snow, the elk population will not suffer. The cows, on the other hand, need all the advantages they can get since all through the long cold winter they will be carrying the calves that are the future of the population.

Winter is a critical season for elk in western Canada. Here, at the northern limit of their natural range, a long winter of deep snow and extreme cold can result in losses of elk to starvation, exposure, and predators. Most elk migrate to areas where winter snow remains shallow. In the prairie and foothills regions, these are generally the few river valleys, hillsides, and ridgetops that are exposed to wind and sun. In the heavy-snow areas of British Columbia, elk migrate to old-growth forests whose canopies intercept a lot of the snow that would otherwise carpet the ground.

Unfortunately, elk winter range is increasingly being used for other purposes. For decades, elk in the East Kootenays and Alberta's southern foothills have had to survive the winter on what forage was left behind by domestic cows. On Vancouver Island and other elk ranges of British Columbia, clearcut logging has eliminated much of the old-growth low-elevation forest that the elk need for winter survival. In Kananaskis Country a golf resort and other tourism developments have replaced one of Alberta's major elk winter ranges. Rural subdivisions, roads, mines, and other forms of industrial development continue to whittle away at western Canada's critical winter ranges.

Nonetheless, elk continue to survive in much of their traditional western range. A series of mild winters in the 1980s, allowed some elk populations to build up to high levels. In Banff National Park the authorities arranged for the Trans-Canada Highway to be enclosed with a special fence as part of a project to widen the highway. Elk deaths due to collisions on the highway dropped from almost 200 a year to fewer than 50, and the elk population tripled in the 1980s.

Elk calves are born in late May and early June. Cow elk seclude themselves on river islands, steep slopes, groves of spruce, or other safe places to give birth to their wobbly legged calves. For the first week or two of a calf's life, it is extremely vulnerable to grizzlies, black bears, wolves, and even coyotes. During early June, cow elk can become very dangerous as they defend their calves from any threat. In recent years several tourists have been badly hurt by cow elk in Banff and Jasper national parks.

By late June the calves are strong and fleet-footed. Bands of cows and calves fatten each evening and morning on the lush new greenery of the northern summer. Bull elk, which shed their antlers in late February, remain separate from the nursery herds, their fuzzy new antlers growing rapidly. All through the glory of the brief northern summer, elk grow sleek and fat on the season's green wealth, preparing for yet another fall rut and long winter. The great cycle of life goes on.

Caribou

WESTERN CANADA'S caribou are in trouble. Where wardens in Glacier National Park once saw herds that numbered in the hundreds, now there are none. In the foothills east of Jasper National Park, caribou populations have crashed to less than one-tenth of their former size. In the highlands of Wells Gray park, the southern Selkirk Mountains, and northern Saskatchewan and Manitoba, the shy woodland caribou are disappearing.

Caribou are wilderness creatures. They prefer the deep snow country of dark northern forests where their wide feet allow them to travel in snow that would bog down a deer or elk. They feed on lichens that hang like hair from old-growth spruce and fir, or on coral-like clumps of caribou lichen that grow on the forest floor in northern pinewoods.

In mountainous areas, caribou migrate from alpine ridges and timberline meadows down into old forests at lower elevations every spring and fall. In spring, they escape the high mountain snowpack during the brief season when it becomes soft and slushy. In fall, the migration is to avoid the soft early snows and find the last greenery of the year.

In the foothills and northern forests there is far less snow. Here, caribou live a more nomadic lifestyle, wandering through unbroken forests to the pine ridges and spruce muskegs where lichens abound.

Logging and other resource development are the biggest problem facing the caribou. As roads and clearcuts open up the landscape, they eliminate the old forests that caribou rely upon. They also expose caribou to increased poaching and legal hunting pressure. As shrubs and young trees fill in the clearcuts, deer and moose populations increase. In response, wolf numbers increase too. In the fragmented habitat left behind by the loggers, wolves find it easier to kill the few caribou than the many deer. Biologists call this kind of a situation a predator pit.

The situation is worst in British Columbia where the most important spring and fall caribou habitats – old forests of cedar and hemlock along the bottoms of mountain valleys – are the most attractive to lumber companies. Little healthy caribou habitat survives in the area surrounding Mt. Revelstoke and Glacier national parks. All of what survives will soon be cut.

There may be other reasons for the decline of caribou. Caribou are adapted to cold climates with long winters. Since the late 1800s, the world's climate has been growing gradually warmer. Hot summers are more common than they used to be, and winter weather is more volatile than it was. It may be that caribou at the southern edge of their natural range are living under such ecological stress that industrial activity is all it takes to push them into decline.

Interestingly, Jasper National Park's caribou populations have in-

They prefer the deep snow country of dark northern forests where their wide feet allow them to travel in snow that would bog down a deer or elk.

creased over the past decade in spite of wolves and other predators. It may be that we can yet save at least some populations of western Canada's beautiful woodland caribou, if we can find it in our hearts to set aside protected wilderness for them.

Mule Deer

THE MULE DEER is the most widespread western deer. Prairie coulees, parkland, aspen groves, northern river valleys, alpine meadows, and sage flats are home to this large deer with its oversized ears and white rump patch.

Unlike the smaller whitetail, mule deer have a distinctive four-legged bounce that they use when frightened. A startled mule deer often looks like it is on a pogo stick as it bounds a few dozen yards and then stops to stare back. Some biologists speculate that this gait evolved because of the mule deer's preference for hilly terrain and log-strewn old burns where the stretched-out running style of a white-tailed deer is less effective.

Mule deer prefer countryside that has some ups and downs. Even so, herds of mule deer often appear far out on the flat grainfields and native prairie of southern Saskatchewan and Alberta. Along the rim of the Milk River canyon, huge mule deer bucks often bed down in the middle of the prairie where they can see danger far before it arrives, and choose an escape route down any of several draws that drop into the canyon.

Mule deer feed mostly at dawn and dusk along the edges of forests or in open, shrubby areas. In fall and early spring, large herds of mule deer can often be seen feeding on the edges of hay fields. During the summer and early autumn, bands of does, fawns, and yearlings prefer aspen forests, grassy hillsides, and the rich vegetation along streams and sheltered slopes. Bucks lead more solitary lives. They may range the high timberline country of the mountains, the foothill forests, or the badlands of the southern prairies.

In November, the elusive bucks suddenly become visible as they cruise the slopes and valleys, necks swollen and antlers newly polished, looking for does. During the November rutting season bucks wander endlessly, joining nursery groups briefly before moving on.

The fawns are born the following May or June. The does normally give birth to two fawns except on poor quality range where they may have single babies.

Mule deer are important prey for cougar, coyotes, and wolves.

During the 1960s and 1970s, mule deer became uncommon over much of their traditional range in Saskatchewan, Alberta, and British Columbia, even as agricultural development and logging allowed them to expand their range north. Because of their open-country habitat and their tendency to stop in the open when startled, mule deer are vulnerable to hunters. More conservative hunting regulations and mild winters have allowed mule deer populations to recover. Now some ranchers consider them pests, due to their habit of breaking into haystacks in winter and trampling the hay they don't eat.

White-tailed Deer

NO LARGE ANIMAL has thrived nearly so well over the past century as the white-tailed deer.

When the first settlers were establishing themselves across the prairie provinces and southern British Columbia a century ago, white-tailed deer were rare. In most areas, they simply did not exist. Today, they are found throughout western Canada east of the Fraser Canyon and south of the Peace River.

Agricultural settlement put an end to the fires that once burned regularly across the northern prairies. Without fire, the willows and poplars that surrounded parkland potholes were able to grow and mature, and they spread by root suckering out onto the surrounding uplands. Gradually, over the past century, what was once prairie with small groves of aspen has become parkland, with large stands of old aspens separated by fields.

In the 1950s, new strains of alfalfa were developed that allowed western Canadian farmers to establish perennial pastures of alfalfa. The combination of maturing forests for cover and winter forage, and rich fields of alfalfa and other crops, created ideal conditions for a white-tailed deer population explosion across the prairies. Since the whitetail is far more elusive and nocturnal than the mule deer, the invasion of the white-tailed deer was helped along by the increasing hunting pressure on mule deer.

The spread of poplars and shrubs along southern Alberta's leaky irrigation ditches provided whitetails with new habitat in what had once been wide-open prairie. In southern British Columbia, agricultural fields, irrigation ditch shrubbery, and heavy hunting pressure on the mule deer also allowed whitetails to expand their range. Massive wolf control programs in the 1950s and early 1960s removed some of the controls on deer populations along the western and northern edges of their range.

Today, the graceful and popular white-tailed deer lives inside the city limits of every prairie city, wanders on southern farmlands, and feeds beside busy highways in the mountain national parks. It has also become a common sight in the southern Okanagan and the Nicola valleys of British Columbia, where it was once unknown. Only the high mountains and the coastlands of British Columbia lack whitetails today.

However, deer are not adapted to northern winters. A cold winter with deep snow can knock back whitetail populations substantially, especially where there is little forest cover. Fortunately, winters have been mild across most of the west in the past decade.

While the changes wrought by the development of the Canadian West have not been kind to animals like the bison, caribou, grizzly, and otter, the white-tailed deer is comfortably at home with the twentieth century. If there is a joker in the deck, it may be in the recent spread of game ranching across the west. Diseases like tuber-

If there is a joker in the deck, it may be in the recent spread of game ranching across the west.

culosis and meningeal brainworm have broken out in these crowded, artificial environments. If infected game ranch animals continue to escape into the wild, the effect on western Canada's wild ungulate populations – deer, elk, and moose – may prove devastating.

Mountain Goat

NO ANIMAL IS as characteristic of the rugged, unforgiving mountain environment as the mountain goat. Long after the elk and mule deer have retreated to their low-elevation winter ranges and the bighorn sheep have gathered on their wind-blown ridges and slopes, the mountain goat remains in its snowy, mountain retreats.

Goats are browsers, not grazers; they eat the twigs, buds, and leaves of a variety of shrubs and small trees. In the winter, mountain goats often remain in deep snow country, banded together on steep slopes and cliffs where they can paw through the snow and find the shrubby forage that they need. Through the long, hard winter, the goats gradually lose weight as the effort of keeping their bodies warm and moving about in the deep snow burns more body energy than their sparse winter diet can replace.

By late April, when the spring avalanches again begin to boom amid the peaks, mountain goats may have lost more than a fifth of their total body weight. With spring, however, the emerging greenery heralds the season of wealth and plenty, and goats rapidly recover the fat reserves they lost during the alpine winter.

High on their rocky slopes, mountain goats need not worry about many predators. Wolves, cougars, coyotes, and bears rarely bother the big white animals. Golden eagles occasionally harass goats and try to knock the animals off cliffs or carry away kids. For the most part, however, the greatest danger facing the mountain goat is slipping on icy cliffs or getting carried away by an avalanche.

Perhaps because they are exposed to less predation than other large ungulates, mountain goats have low reproductive rates too. A nanny is generally at least three years old before she has her first kid, and then she has only one per year. Baby mountain goats are able to walk the day they are born and follow their mothers over the most precipitous terrain, with little risk of injury. From a distance, it is often hard to tell whether a nanny has a kid, since the babies stay beside their mother, generally on the uphill side.

Mountain goats have a taste for mineral soil, especially during the weeks of spring moult when they shed heavy coats of white wool. In June and occasionally throughout summer, nursery herds of nannies, kids, and yearlings visit mineral licks along the Banff-Jasper and Banff-Windermere highways and along the Trans-Canada Highway snowsheds near Glacier .

Bighorn Sheep

FROM THE WINDSWEPT ridges of the Rocky Mountain front ranges to the brown slopes of the Fraser Canyon, bighorn sheep choose the most dramatic of landscapes as their dwelling places.

Short-legged, built for climbing rather than running, bighorns are rarely found far from steep slopes where they can retreat from predators. In fact, most of the sheep that I have found killed by wolves were along valley bottoms and on broad alpine meadows. They had wandered too far in search of grass and been cut off from their escape to nearby cliffs.

The short legs of bighorn sheep limit where they can range. Any place where the winter snow gets more than a foot deep is no place for sheep. Even if they fail to get bogged down, the energy bighorns waste in plowing through the drifts will soon exhaust them and set them on the long downward spiral of winter starvation. Sheep are forced to subsist on dry, brown grass throughout the long mountain winter. They are engaged in a long, slow race against starvation, with no margin for error. A winter range for bighorn sheep needs steep slopes, grass, and little or no snow.

Ironically, one of the snowiest places in the Canadian Rockies is also the winter home of some of the biggest sheep in Jasper National Park. The Columbia Icefield area gets up to 10 metres or more of snow each winter, yet a band of up to 50 sheep spends each winter across the valley on Wilcox and Tangle ridges. The vast sweep of the Icefield to the west spills strong, reliable winds eastward, sweeping snow from the grassy slopes all through the winter. Farther east, where winds funnel through Front Range valleys or over the Alberta foothills, still larger herds of sheep return each winter to feed on the windswept alpine grasslands. In British Columbia, most sheep migrate to low elevation grasslands where mild temperatures, rather than wind, keep winter snow accumulations light.

In spring, bands of sheep follow the retreating snow back into the mountains seeking out the freshest and most nutritious greenery. All during the summer bands of bighorn rams fatten in remote valleys and mountain slopes while nursery bands of ewes and lambs travel and feed among the high basins.

Bighorn sheep have an elaborate social structure that helps them survive in an environment where their summer and winter ranges are often highly fragmented and far apart from one another. Older animals dominate and lead the younger animals, passing on traditions of travel and behaviour that are essential to the survival of the population. One of the most dramatic ways that bighorns establish their social order occurs among the ram herds. Similar-sized rams test each other by rearing up on their hind legs and then crashing forward against each other. After each clash the two

Any place where the winter snow gets more than a foot deep is no place for sheep.

rams turn their heads sideways to one another to display their horns. The message appears to be: "Next time you see these, remember that headache you just got, and back off !"

Pronghorn

IN THE SHIMMERING heat of a prairie afternoon, the first sign of pronghorns is the sudden flash of white amid the low sage and speargrass. Then the prairie appears to fragment, and a dozen richly-patterned animals race across the plain only to stop abruptly, a few hundred metres away, and stare back.

Pronghorns are strikingly-patterned animals, yet they seem to disappear against the dusty greens and browns of the prairie grassland when they stand still. At the first sign of danger, pronghorns flare the white hairs of their large rump patches, semaphoring a danger signal to other pronghorns in the vicinity.

Prairie is often described as flat. In fact, most of the western plains are rolling, with low drainage channels winding between gently rising uplands. The pronghorns that roam the shortgrass plains know every dip and rise of their home ranges, and they use the landscape to their advantage. At the first hint of danger, herds of pronghorn race away at speeds approaching 85 kilometre per hour, disappearing into a seemingly featureless plain.

In Canada, pronghorns are at the northernmost extent of their natural range, and they are not adapted to life in snow. As a consequence, severe winters control pronghorn populations. Heavy snowfall, especially when accompanied by winds that build deep snowdrifts, can trap herds of pronghorn and lead to starvation or death from exposure. During the severe winter of 1977-78, thousands of pronghorns died on the southern Alberta and Saskatchewan plains. Winter kills are aggravated by roads, fences, and railroads that trap snow and create long barriers that block the animals from wandering to more hospitable areas. Whole herds of pronghorn are sometimes killed by trains when they gather on wind-exposed railway grades to escape the drifts.

In spite of the setbacks they suffer during severe winters, pronghorns are relatively common and widespread in the shortgrass prairies of southern Alberta and Saskatchewan. Each spring, the does give birth to twin fawns. At first, the fawns avoid danger by lying flat on the ground while their mothers feed. However, within less than two weeks, the fawns are nearly as fleet of foot as their mothers. The dangerous time for fawns is the period before they can flee from enemies; this is when many baby pronghorns fall prey to bobcats, coyotes, and eagles.

Pronghorns feed on herbs, shrubs, and other plants. They get most of their moisture requirements from food, rather than from drinking. This is essential, since they occupy the driest country in the west. In winter, the sagebrush flats along coulee slopes and river bottoms are critical habitat for pronghorns, as are alfalfa fields and other places where vegetation reaches above the snow.

Regulated hunting plays an important role in keeping pronghorn populations in balance with their prai-

At the first hint of danger, herds of pronghorn race away at speeds approaching 85 kilometre per hour, disappearing into a seemingly featureless plain.

rie habitat. This is essential today, now that fences, roads, railways, and other developments prevent the long, free-ranging migrations that pronghorns formerly made to winter, spring, and other seasonal ranges.

Bison

SOUTH OF CALGARY, where the brown Porcupine Hills pinch out against the Oldman River valley, piles of lichen-encrusted rock lie amid the prairie grasses. High above the plains, the endless chinook wind hisses and whines as it hunts among the prairie vegetation, but the land is empty. Only the ghosts of the bison still wander here.

For many centuries, these rock piles were part of an elaborate pattern of markers and decoys used to hunt the herds of plains bison that wintered among the grassy hills. Far out on the plains, the prairie grass lay beneath drifts of crusted snow, but here where the chinook wind roars almost constantly, the brown hills were usually bare. Native people camped in teepees among the cottonwoods of the Oldman River and hunted the bison on the flanks of the hills.

A few men would show themselves to the grazing animals and start them drifting away. Other men would appear, turning the animals. Some of the rock piles were set up to look like yet more human figures, so that the bison soon began to feel hemmed in. As they drifted ahead of the drivers, their speed increased, until finally they began to run to get ahead of their pursuers.

However, the drive lanes were intelligently designed to take advantage of the rolling landscape, and each hunt was carefully planned to take the wind and the location of the herds into account. After a time, in one last burst of pursuit, the Natives forced the fleeing bison over a hill that concealed a sandstone cliff, and before the herd could turn, many of the great animals went crashing into the tangled brush below.

Then the work began, for the bison was the Plains Natives' survival; it provided food, hides for clothing and shelter, and bones for tools.

The great buffalo hunts are only forlorn rumors on the empty chinook winds today; the bison are gone. At Head-Smashed-In Buffalo Jump the story is still told by the descendants of those early hunters, but only bones and stone tools remain of the bison that once roamed here. Farther north, above Alberta's Red Deer River, another great hunting spot is commemorated at Dry Island Buffalo Jump. In fact, there were many buffalo pounds, as the old kill sites were called, along the creeks, rivers, and coulees of the southern prairies.

The extermination of the plains bison occurred with such speed, and on such a scale, that it seems incomprehensible. In 1800, there are estimated to have been 60 million buffalo on the Great Plains. By 1850, there were only half as many. By 1890, the bison were gone; fewer than 50 survived in the world. The prairie was littered with bones, the native people of the plains were reduced to poverty-stricken, haunted survivors, and settlers were flooding west to convert the prairie wilderness to farmland.

Far to the north, the lankier wood bison survived in the boreal forests and wetlands of the Mackenzie watershed.

Today, plains bison are raised as specialty livestock by some ranchers. Display herds live confined lives in

In 1800, there are estimated to have been 60 million buffalo on the Great Plains. By 1890, the bison were gone; fewer than 50 survived in the world.

national parks. The endangered wood bison is gradually being re-established in viable herds in northern Canada.

But the plains and foothills are full of ghosts, the wind hunts endlessly across the tamed farmlands of prairie Canada for the scattered remnants of native grassland, and nobody will ever again hear the thunder of a million hooves as the great brown living tide sweeps across the endless, open plains.

Pikas & Hares

THERE IS A RABBIT for all of western Canada's habitats: from coast to mountaintop and from northern forest to dry prairie.

American Pika

Few animals are more delightful for the mountain hiker to watch than these busy little rock rabbits. It might seem strange to think of pikas as rabbits, since their small ears, loud calls,

and scurrying behaviour might seem inconsistent with most people's idea of a bunny. However, rabbits they are.

Pikas are rock-pile specialists; they occupy dens deep within the rocky rubble that accumulates in large talus fans at the foot of mountain cliffs. Along Banff's Vermilion Lakes and in other spots along the Trans-Canada, Yellowhead, and Crowsnest highways, pikas have taken advantage of the work of highway engineers, colonizing the rock fills beside highways. But for the most part, this robin-sized rabbit is restricted to the higher valleys of the western mountains.

Unlike most wild animals, the pika is most active during the daytime. Even so, most hikers hear pikas before they see them, because they blend so well with their rocky surroundings. Pikas alert one another to danger with a distinctive braying beep.

The subalpine world of the pika produces little in the way of food, and pikas busy themselves all day long feeding on green foliage, scurrying to and from patches of grass and flowers. Since pikas do not hibernate, they also need winter food supplies and have the unique habit of storing their own winter hay. Hikers commonly see bits of grass, flowers, herbs, and other vegetation carefully laid out to dry on boulders. Once the harvest is well enough cured, the pika will add it to a growing haystack deep in the boulder pile.

When the alpine world is draped with deep snow and the mountains seem cold and lifeless, these little bun-

dles of life will be active beneath the snowpack. They will be scurrying about amid their rocky rubble, nibbling on the hay they stashed away during the brief green wealth of their mountain summer. Cross-country skiers trace their way through a frozen world, unaware of the busy little rabbits living away the winter in the semi-darkness of the snow-shrouded slopes nearby.

Pikas give birth to three or four babies each year. Weasels are their main predator.

Snowshoe Hare

Pity the poor snowshoe hare when winter comes late.

This hare's unique adaptation to its northern world actually works against it when the snows are delayed. Each year, snowshoe hares go through two colour changes. In April, as the winter snow disappears, they turn brown. In late October, they turn white. For an animal in high demand among northern predators, seasonal camouflage is critical. However, if the winter snows are late, the white hares stand out vividly against the dead vegetation of the November forest, making them easy pickings.

The snowshoe hare has large feet that allow it to travel easily on deep snow. As snow depths increase through the winter, hares are able to feed on bark and twigs at progressively higher levels in the forest. Where there are dense stands of tall willows or young spruce, hare trails can eventually become well-defined, heavily trampled networks through the winter forest.

Snowshoe hares are important prey for a variety of animals. Since they are active almost exclusively at night, they need to fear nocturnal hunters. Lynx, in particular, depend on these northern hares for food. Great-horned owls, weasels, coyotes, foxes, and other predators also hunt them, as do humans. Fortunately, the snowshoe hare is quite prolific. A female may produce up to four litters of babies each year, giving birth ultimately to as many as 20 babies in a season.

The snowshoe hare is found throughout the forested areas of Canada, except for Vancouver Island and some other coastal areas. Their numbers vary considerably from year to year, and populations are quite cyclical. Every decade or so the hare population builds up to the point where there seems to be one hare under every bush; but the population crashes the following year. It appears that starvation is the main factor that causes populations to crash, although disease can also be important. Predation is what keeps numbers low over the following year or so until the predator populations drop too.

White-tailed Jackrabbit

One day I spotted a coyote sneaking along a fenceline near the Trans-Canada Highway east of Calgary. It eased itself into a clump of tall grass not far away, and lay down.

Another movement caught my eye. Far across the field, another coyote was loping across the shortgrass prairie behind a jackrabbit. Neither the hare nor the coyote was working very hard, but the jackrabbit's path was clearly going to take it right past the waiting coyote.

Closer and closer they came, until I could see the big jackrabbit's protruding eyes and long muscular legs propelling it into each effortless bound. As it drew even with me, the waiting coyote exploded from the grass, a tawny streak racing to intercept the jackrabbit.

The coyote never had a chance. With no apparent effort at all, the jackrabbit veered left and streaked off across the prairie, leaving its two pursuers to trot away, tongues lolling, trying to act as if they hadn't really wanted that jackrabbit anyway. Far out on the prairie the jackrabbit slowed to a stop and sat watching, its long ears erect, as the coyotes disappeared into a coulee.

Where cottontails and snowshoe hares rely on dense cover and camouflage to protect them from predators, jackrabbits rely at least as much on speed. Long, lean, and powerful, a jackrabbit has nothing to fear from a coyote, swift fox, or other predator in a fair chase. Most successful land predators rely on stalking or ambush to catch jackrabbits.

Eagles and hawks are important predators, too. One day I flushed a golden eagle from a fence post beside a prairie stubble field. As it flapped heavily away, it must have spotted a jackrabbit crouched in the field, for it suddenly dropped its talons, then landed and ran a few paces back towards me, like a big awkward turkey. A moment later, a jackrabbit came racing across the field, only to topple over beside the fence, dead. The eagle's razor-sharp claws had inflicted mortal injuries.

White-tailed jackrabbits live only in the prairies. They prefer short native grasses to taller grass or cultivated fields; consequently, they coexist well with the domestic cow. Heavily grazed pastures are good jackrabbit habitat. Jackrabbits will use other shortgrass habitats too; they are commonly seen on the campuses of the Universities of Calgary and Saskatchewan and in large urban parks on the prairies.

Like the snowshoe hare, the jackrabbit changes colour in winter. The fall moult begins in November and goes into December. Some jackrabbits become a pale grey rather than pure white in winter.

Jackrabbits breed once or twice each summer, producing up to ten young per female.

Porcupine

NATIONAL PARK managers learned long ago that plywood does not belong at timberline. Plywood signs, outhouses and window shutters simply don't last; porcupines eat them. One of the better ironies I ever saw was a sign in Yoho National Park stating: "It is unlawful to feed animals." A porcupine was eating the sign, which was already half gone when I saw it.

One outhouse in Kootenay National Park was unusable; a porcupine had eaten out the entire seat and was living in the pit.

Today, most outhouses in national parks are built of fibreglass or solid wood, because the ubiquitous porcupine will not be cured of its taste for the glue that binds plywood. The substitutes for plywood and particle board leave the porcupine free to enjoy its normal diet of buds, twigs, and bark, while leaving hikers free to enjoy undamaged facilities.

Porcupines are widespread in western Canada. Even in the middle of the prairies, a yellow-haired prairie race of porcupines lives comfortably along small streams and coulee slopes, wherever there are shrubs and trees for food. However, the highest density of porcupines occur in the timberline country along the spine of the Rocky Mountains. In an evening hike in the Yoho Valley, I counted more than 20 porcupines.

Such abundance seems even more remarkable when you consider that porcupines have only one baby each spring, a very low reproductive rate. Unlike other rodents, most porcupines live to a comfortable old age without much to worry about from other animals. One experience with a porcupine's quills is generally more than enough to convince a coyote or lynx that other foods are preferable. In fact, porcupine quills are designed to work themselves into the body where they can cause infection or pierce internal organs. The placid, slow-moving, heavily armored porcupine is not be trifled with.

Even so, some predators are adept at hunting and killing porcupines. A successful strategy for a fisher, wolverine, or other carnivore is to prevent the porcupine from hiding its face against a tree or log, as they normally do, and then to dart about until it can bite the porcupine's snout. A porcupine's belly is unprotected by quills, making it vulnerable to attack by any predator that can turn it over.

Porcupines were once considered vermin because of their propensity for chewing on leather, axe handles, and other salty surfaces. One old-time warden in Jasper National Park reported killing 14 porcupines in one night at Wolverine Cabin, because they persisted in chewing on the cabin door. Today, porcupines are carefully protected. The distribution of porcupines in the western mountains corresponds strongly with the distribution of wolverines. It may be that the

One experience with a porcupine's quills is generally more than enough to convince a coyote or lynx that other foods are preferable.

porcupine plays a vital role as the only medium-sized prey available in mid-winter in the deep snow of the high mountains.

Woodchuck

Because woodchucks feed on lush vegetation and prefer dense cover near their burrows, they generally avoid over-grazed pastures. Abandoned woodchuck dens are favourite homes for foxes and other small animals.

A fat woodchuck is a windfall for most predators, and it is fortunate that woodchucks are such prolific breeders. A female woodchuck will produce up to six or more babies each summer. Young woodchucks disperse in the autumn, searching for their own

THERE ARE PROBABLY more woodchucks in western Canada than there were a century ago. Woodchucks thrive in meadows and farm fields along the fringes of forested country. They are most common in the aspen parkland region of the prairie provinces, although a western subspecies occupies forest openings through the upper Fraser and Columbia watersheds in British Columbia. Agricultural clearing in these areas has produced ideal habitat for woodchucks, and predator control has improved the odds of survival.

Woodchucks are constantly doing home-improvement work. They dig elaborate burrow systems that typically have several entrances covered with fresh soil. From the burrow mouths, trampled runways fan out into the surrounding meadows. Woodchucks graze on green vegetation, keeping a wary eye open all the time for red-tailed hawks, foxes, coyotes, and other predators.

home meadows, and they are particularly vulnerable. From the predators' point of view, the autumn abundance of fat young woodchucks is fortuitous. Hawks need to accumulate fat reserves to help with their migration flights while coyotes and other earth-bound predators are facing a long, hard winter and have a similar need.

In dry rocky areas along prairie rivers, in the southern parts of the Alberta foothills, and in the interior valleys of British Columbia, the yellow-bellied marmot fills a similar ecological niche to the more northerly woodchuck. On Vancouver Island, the endangered Vancouver Island marmot occupies high-elevation meadows surrounded by clearcuts and forest relics, living a life similar to the woodchuck of the mainland.

Woodchucks ... dig elaborate burrow systems that typically have several entrances covered with fresh soil.

Hoary Marmot

THE HOARY MARMOT is a timberline equivalent to the woodchuck. Marmots are large animals, however, sometimes more than twice the size of the low-elevation woodchuck.

Hoary marmots are daylight animals that feed on the rich herb meadows that grow along mountain streams and amid the boulders along the edges of high valleys. They like to sun themselves on the tops of rocks. Long before a hiker sees a marmot, the distinctive loud warning whistle will give away its presence. Marmots generally allow people to approach them, since they need only clamber down from their sentry boulders to find shelter in their underground den.

Because marmots den in rocky areas along the edges of high mountain meadows, they rarely need to fear predators while they are at home. Of all the high-country predators, only the grizzly is adapted for digging food. But they prefer to dig for legume roots and grubs than to invest the energy required to excavate a marmot from its bouldery den.

Above ground, however, the marmot is at risk from a variety of predators. Golden eagles, coyotes, wolverines, and other mountain predators all hunt the marmot, because so large a rodent is a windfall compared to the smaller voles and mice that are more abundant and easy to catch.

Hoary marmots give birth to four or five babies late in the spring. Perhaps because of the short growing

season, it takes them two years to grow to full adult size.

Marmots hibernate in winter. Unlike bears, which merely go into a deep sleep, a marmot's body temperature drops extremely low and its metabolism nearly stops. Bears have to burn fat to maintain their body processes, whereas marmots conserve energy by shutting down their systems. In spring, with the disappearance of the snowpack and the appearance of the first green vegetation, marmots emerge from their underground dens and resume their bucolic lifestyles for another green, bright summer beneath the high peaks of the western mountains.

A distant relative, the Vancouver Island Marmot, is more similar to the woodchuck. It lives in lush meadows in forests at high elevations and is considered endangered.

Squirrels & Chipmunks

NATIVE SPECIES OF squirrels are readily seen just about everywhere in the West. In the past two decades both the black and grey varieties of the eastern grey squirrel have expanded their range to the cities of western Canada.

Richardson's Ground Squirrel

Every prairie child knows the pale, sand-coloured Richardson's ground squirrel, or "gopher" as it is more commonly known. Along with the buzz of the clay-coloured sparrow, the rattle of grasshoppers, and the smell of pasture sage, the high-pitched squeak of the gopher is a characteristic part of every prairie summer.

Richardson's ground squirrels prefer short grass to tall. They often form loose colonies, and there are usually one or more ground squirrels on the watch for danger at any given time during the day. From a distance, their erect posture makes them look like stakes stuck in the prairie, giving rise to one of their popular names, "picket-pin." They thrive in heavily grazed pastures throughout the prairie region, as well as the edges of grainfields, golf courses, and lawns. In fact, they thrive so well and eat so much grass and grain, that they are generally considered agricultural nuisances. Over the years these little rodents have been subject to a wide range of persecutions – from bounty hunters, poisoning campaigns, and target shooters – but still they thrive.

In part, their success is due to their capacity for reproduction. A female may give birth to as many as eleven babies each year. In late summer the young disperse and are vulnerable to road-kill as well as to predation by Swainson's and rough-legged hawks, prairie falcons, coyotes, and others. One predator that specializes in hunting and killing these prairie rodents is the long-tailed weasel. Without the abundant Richardson's ground squirrel, the prairie world would be a poorer place, with far fewer of the dramatic predators that rely on them.

In fact, successful poisoning campaigns have kept ground squirrel populations so low in the Milk River area of southeastern Alberta that ferruginous hawks and prairie falcons are uncommon. Farther north, in the Red Deer River watershed, hawks and falcons are more common, perhaps because of the land-use history of the area. Much of the Red Deer River country was farmland prior to the Dirty Thirties when it was abandoned. The damaged soils have been gradually restored to rangeland, but the sparse vegetation cover is ideal for the Richardson's ground squirrel. And abundant ground squirrels support abundant hawks.

Ground squirrels are true hibernators. While a few Richardson's ground squirrels may remain above ground through the Indian summer days of October, most vanish below ground in late August, not to be seen again until late in March. The earliest ground squirrels to leave their winter dens often burrow up through drifts of late winter snow. Exposed on top of the snow drifts, the early risers are easy prey for migrating bald and golden eagles who move north along the foothills and western plains at this season.

Black-tailed Prairie Dog

Larger than the common Richardson's ground squirrel, the prairie dog also lives in colonies. In the late 1800s, huge colonies of these sand-coloured rodents covered the shortgrass plains of North America, including parts of southern Saskatchewan.

To the rangemen who were settling the prairies, any native creature that ate grass was vermin, especially if that creature dug holes that could break the leg of a horse or cow. The colonial nature of prairie dogs made them highly vulnerable to the poisoning campaign that drastically reduced their numbers around the turn of the century.

In Canada, prairie dogs were nearly extirpated. A few small colo-

Richardson's
Ground Squirrel

Black-tailed
Prairie Dog

protect them from being flooded by the rare torrential summer rains. Sometimes, instead of a prairie dog, the erect figure on the mound at the entrance to an old prairie dog den may be a burrowing owl. They are unique prairie birds that nest underground in abandoned burrows.

In Canada, prairie dogs are in danger of being killed and eaten by badgers, coyotes, weasels, rattlesnakes, and hawks. And farther south, the black-footed ferret specializes in hunting prairie dogs. Nearly extinct, ferret populations are being slowly nursed back to health by US biologists; some day Canada's prairie dogs may again encounter their traditional foes in the windswept prairie wilderness of the Grasslands National Park.

nies survived in the vicinity of Val Marie, Saskatchewan. Even so, one of these was lost to flooding by an irrigation dam, and others were poisoned out by area ranchers.

Today, Canada's only surviving prairie dog population is protected by the Grasslands National Park in the Frenchman River watershed near Val Marie. The park was established in the nick of time, since one colony was poisoned out just before the agreement for the park was finally signed.

Prairie dogs are highly social animals, and a visit to a dog town can be interesting experience. They can be found barking and scurrying, chasing one another, dust-bathing, sunning themselves, and grooming or nuzzling one another. Raised mounds of excavated dirt at the mouths of burrows

Columbian Ground Squirrel

Larger and more richly coloured than its prairie cousin, the Richardson's, the Columbian ground squirrel is the common grassland squirrel of the Rocky and Columbia mountains. Campers in the mountain national parks soon become familiar with the sight of these attractive rodents and with their distinctive peeping calls.

Columbian ground squirrels range from the foothill grasslands of Waterton Lakes National Park and the Bar U Ranch National Historic Site up to the alpine meadows of Jasper and Glacier national parks. They prefer grassland but also frequent recently burned areas, heather meadows, and other open habitats. In British Columbia,

Columbian ground squirrels range into the interior valleys where they occupy grassland and sagebrush areas.

Female ground squirrels may give birth to as many as nine babies each spring. The young squirrels remain in their underground burrows until midsummer, dispersing late in August. Once out of their dens, the young ground squirrels face the constant threat of death from eagles, hawks, coyotes, weasels, and most other mountain predators. Columbian ground squirrels, being widespread, common and relatively large, are prey items for many other animals.

Like other ground squirrels, this species is a true hibernator. Curiously enough, the ground squirrels that live at low elevations go to bed earlier each fall than those living near timberline. This appears to be related to the fact that low-elevation ground squirrels get an earlier start each spring and can reproduce and fatten up earlier in the year than the higher elevation animals who have to wait longer for the spring snow to melt.

Ground squirrels are readily tamed, and in popular recreational areas they learn to beg for handouts from tourists. In general, feeding ground squirrels is not a good practice since it makes them vulnerable to predators and, potentially, to digestive problems. The ground squirrels can be dangerous to humans, too, because they have been known to bite the hand that feeds them. They also carry ticks and other parasites.

One little-known danger of feeding ground squirrels is that it makes them vulnerable to vandals. Several families of Columbian ground squirrels in Kootenay National Park were beaten to death with sticks by a family of young boys whose rural background had taught them to consider ground squirrels as vermin. If the animals had not been used to seeking handouts from humans, they would not have died.

Columbian Ground Squirrel

Golden-mantled Ground Squirrel

Most people who see this rodent for the first time assume they are looking at a big chipmunk. However, the golden-mantled ground squirrel is a different creature. Its chipmunk-like back stripes only extend as far as its neck, rather than continuing through its eyes. It has a rich golden-brown nape, and a whitish ring around its eyes.

Golden-mantled ground squirrels like rocky, open country with a scattering of trees. In the Rocky Mountains they are commonly seen along the edges of canyons and mountainside trails. Farther west, in British Columbia, they occupy the edges of dry forests and steep rocky slopes. As often as not, the same habitat that contains these striped ground squir-rels may also support true chipmunks, Columbian ground squirrels, or marmots.

Golden-mantled ground squirrels eat a variety of vegetation, from the leaves of grasses and other plants to flowers, seeds, and roots. The underground fruiting bodies of fungus can be important food sources at times, as they are for both the red squirrel and the flying squirrel. Golden-mantled ground squirrels cache food late in the summer, with the inevitable result that they help flowering plants and fungi to spread and regenerate themselves.

Like most other ground squirrels, females give birth to four to eight babies each spring. Golden-mantled ground squirrels are rarely as common as other species, but even so they are important prey for hawks, weasels, coyotes, and others.

Least Chipmunk

The least chipmunk is the common chipmunk of the forested parts of the prairie provinces, northern British Columbia, and the Rocky Mountains in Alberta. From the Rockies west across southern British Columbia, the yellow pine chipmunk, virtually identical to the least chipmunk, takes over. Townsend's chipmunk, a slightly larger species, has been introduced to southern Vancouver Island.

Chipmunks are small, nervous animals that frequent forest edges, tangles of brush and deadfall, and the shrubby areas along streams and roadsides. Their seeming nervousness is warranted; most predators feed on the chipmunk. However, chipmunks live less-exposed lives than ground squirrels, since they spend much of their time rummaging about in tangles of shrubbery.

Chipmunks eat a wide variety of food: insects, grubs, seeds, nuts, leaves, and buds. Late in the summer, when fruit and berries are ripe, the signs of chipmunk activity are easy to detect. Chipmunks eat the seeds of rose hips, buffaloberry, saskatoons, and other fruit. They leave the flesh and skin behind, usually in tidy piles on top of a rock, log, or stump.

Like most small rodents the chipmunk is an important prey item for hawks, weasels, and others. And, like most small rodents, the chipmunk is a prolific breeder. A female chipmunk will have one litter of up to six or seven babies each year.

Chipmunks hibernate in winter, awakening periodically in their under-ground dens beneath the snow to feed on the food they stored late the previous summer.

Red Squirrel

It is nearly impossible to visit western Canada without encountering the red squirrel. Red squirrels prefer the evergreen forests of the northern prairie provinces and the western mountains, but they also live in the aspen parkland region and in well-forested parts of some cities and towns. Over the last few decades, as urban forests have matured, the larger, introduced eastern gray squirrel has also become established in western cities.

Red squirrels are both widespread and common. They specialize in feeding on seeds from the cones of coniferous trees, as well as the mushrooms and underground fruiting structures of the many kinds of soil fungus that occur in coniferous forests. Dense, dark forests of spruce, fir, and pine

Least
Chipmunk

produce little in the way of edible material for most mammals, but the red squirrel's specialized feeding habits allow it to thrive in this habitat.

Considering the fact that goshawks, owls, martens, fishers, and other predators are constantly on the lookout for a meal of squirrel, red squirrels are surprisingly cocky. Any animal that wanders within sight of a squirrel is bound to be scolded until it is out of sight.

Squirrels do not hibernate in winter; instead they lay in middens – vast stores of pine, spruce, or fir cones. The stashes of cones are used year after year, until a substantial pile of cone scales and stems has grown upon the forest floor. Old middens are riddled with burrows where the squirrels take refuge during exceptionally cold or stormy winter weather.

Like other forest rodents, the red squirrel has a large family each year of up to eight babies. Unlike other forest rodents, however, the red squirrel is not just a prey item for larger predators; it kills and eats other animals itself when the opportunity arises. Eggs, mice, and young birds all are part of the red squirrel's diet.

Red Squirrel

Northern Flying Squirrel

One of the most beautiful animals of the northern forest, the flying squirrel is also one of the least often seen. Its huge, soft eyes are adaptations to night living, for this species is almost exclusively nocturnal. Those soft thumps and scuttling noises on the roof of the cabin or mobile home late at night are probably made by this small forest squirrel.

The flying squirrel is found throughout the forested regions of the West, preferring old forests with at least some coniferous trees mixed in. The unique flaps of skin joining each foreleg with the corresponding hind leg combine with the flying squirrel's broad, flat tail to allow it to glide from one tree to another through the darkened forest.

We who tend to segregate the world into separate, identifiable things have a lot to learn from the ecology of the flying squirrel. The flying squirrel is so intimately connected to the old-growth forests that it is fair to ask whether there really is a point at which the forest ends and the squirrel begins.

The flying squirrel nests in the old-growth trees. And although flying squirrels eat fruit, lichens, seeds, and other vegetable matter, one of the major items in their diet is the underground fruiting bodies of fungi. Many of these fungi are essential to the survival of forest trees, because strands of fungus coat the ends of tree roots, providing the trees with nitrates, water, and other necessary elements of life. In exchange, the fungi benefit from the trees by sharing in the sugars and starches produced in the leaves and carried to the roots in sap.

Without the fungus, the trees would not thrive. Without the trees, the fungus would not thrive. And without the trees and the fungus, the flying squirrels would not thrive.

Flying squirrel droppings contain organic material, bacteria, and fungus spores. As the flying squirrel travels through the forest, it inoculates the forest with the fungi that the trees need. In the old-growth forest ecosystem, then, as in every ecosystem, the things we view as being different and unique from one another are actually intimately connected. It may be that the old-growth forest ecosystem is more real, as a recognizable entity, than the flying squirrel or the fungus. Just as a heart cannot exist without a body, and a body cannot exist without a heart, the forest is more than the sum of, and is intimately dependent upon, each of its component parts, including the flying squirrel.

The flying squirrel is far more common than it seems. Flying squirrels produce about three young each year in their nests of tangled lichens, moss, and bark. Their major enemies are owls, martens, weasels, and large clearcuts. However, clearcuts are the most dangerous, since animal populations can almost always recover from predation. Habitat loss is incurable.

Beaver

MANY OF THE highway routes across the prairies into the Rockies and on to the Pacific coast parallel the water and overland routes of the early fur trade brigades. Rocky Mountain House, Jasper, Invermere, Prince George, Fort St. James and many other western towns and historic sites commemorate the age of exploration and adventure that was spawned by European demand for beaver pelts. The beaver adorns the Canadian five-cent coin and the Canadian Parks Service logo.

The beaver is an animal, like man, that re-shapes the environment to meet its needs for life. Before the arrival of the Europeans, nearly every watercourse in Canada contained beavers. Strings of ponds held back by beaver dams extended up tributary streams to their headwaters. Even streams that only flowed seasonally were plugged by these obsessive dam-builders, creating ponds that lasted through the dry seasons.

The demand for felt hats in Britain during the late 1700s and early 1800s created a market for the unique dense fur of the beaver that was processed, into a rich, lustrous felt. The European beaver was rare, but in Canada the beaver was everywhere. Trappers and traders invaded the interior of the continent, pursuing colonies of beavers up every stream and watershed. As the easternmost populations were exterminated, the adventurers moved west, setting up trade with the Native people whose lands still teemed with beaver.

But by the late 1800s the market for beaver had crashed, coincidentally with the depletion of most accessible beaver populations. Conservation was unheard of and trappers and market hunters severely depleted wildlife populations and moved on to new terrain. And with the shift to railway and other forms of overland transportation, the old fur trade routes and posts fell into disrepair. Canadians looked to new frontiers for their resource profits.

Since the 1930s, when populations reached their lowest point, the beaver has rebuilt its numbers. These large rodents breed once each year, producing up to eight babies, or kits, each spring. Coyotes, wolves, cougars, and other predators hunt beavers, but they are still able to build their populations rapidly under the right conditions.

Today, beavers are found along most prairie creeks, in the heart of major cities like Saskatoon, Regina, Calgary, and Edmonton, and all through the forested regions of the West. Late on a summer evening, people often gather beside the streams and lakes of western Canada to watch beavers cut trees and carry the branches out to underwater caches or to build up their dams and lodges. Like a rifle shot, the crack of a beaver's tail on the surface of the water warns the watchers when they get too close.

The beaver is an animal, like man, that re-shapes the environment to meet its needs for life.

The beaver's death and exploitation helped build this country; its life, today, enriches those of us who are fortunate enough to be here now.

Muskrat

A SMALLER COUSIN to the beaver, muskrats often live in the same ponds and slow-moving streams. Muskrats, unlike beavers, rarely eat the bark of trees or shrubs. Instead, they eat the roots and tubers of water plants, as well as pondweeds, snails, mussels, and other pond life. As a result of their different feeding habits, muskrats are far more widespread than beavers on the prairies. The large sloughs and lakes of the prairies are poor in woody growth but rich in other vegetation.

The muskrat is mostly an evening and nighttime animal. Much smaller than the beaver, it can often be recognized by its habit of lying still in the water with its narrow, ratlike tail lifted partway out of the water at an angle. If danger threatens, the muskrat disappears underwater where its webbed hind feet and powerful legs allow it to swim considerable distances. It may retreat into a bank den or floating lodge.

Although less important than bea-ver pelts, muskrat fur was also important to the fur trade, and muskrat trapping continues to be important to the economy of many northern native communities. On the delta of the Peace and Athabasca rivers, and in other large, shallow sloughs and lake deltas, muskrat populations can be extremely dense. Since females produce up to seven babies per litter, and two litters each year, muskrat populations can explode under the right conditions. In good muskrat habitats the characteristic pushups – small lodges built from vegetation and mud, containing muskrat dens – often dot

the shallow water across wide areas.

Muskrats may be prolific, but they face many enemies. The biggest threat is winter weather and water conditions. If water levels drop too low in winter, muskrat pushups are left high and dry, and their underwater entrances freeze solid. If water levels rise in winter, the pushups are flooded. Ever since BC Hydro built the W.A.C. Bennett dam on the Peace River, muskrat populations on the Peace-Athabasca delta, more than 1600 kilometres downstream, have crashed. River levels are now controlled by computers. And the computers respond to consumer demands for electric power, not to the ecological needs of muskrats.

Where habitat is more reliable, muskrats need only worry about their normal enemies: mink, coyotes, bald eagles, and other predators. Muskrat pushups are maddening to hungry winter coyotes. The mud and vegetation freezes solid. So the poor coyote can smell the muskrats, but cannot dig down and get them. Some prairie sloughs are criss-crossed in winter with the trails of coyotes wandering from one pushup to another ... hoping.

Grizzly Bear

EARLY IN APRIL, as the first thawing winds of spring hiss and whisper in the pines, the first grizzly tracks appear in the granular old snow of the high valleys. Bigger than a human hand, with an arc of clawmarks extending a good five centimeters or more from the toe imprints, the pigeon-toed tracks trace their way in and out of the timber, along the snowdrifted meadows that line newly awakened creeks and streams.

The great bear wakes up hungry. From mid-November, when it settles down in its winter den, until early April when it heads out into the rumor of spring, a grizzly can lose almost a quarter of its body weight. Through the months that follow, a grizzly's life is an endless search for high-energy food – a critically important search, because if a grizzly goes to bed hungry it may never see another April dawn.

Not that long ago, as the mountains measure time, grizzly bears ranged the entire West. From the badlands of the Milk and Frenchman rivers, through the aspen parklands of central Saskatchewan and Manitoba, all the way to the coastal estuaries of British Columbia, the great silver-tipped bear ranged at will, following its nose through the seasons.

In early spring, the great bears rooted for legumes on dry slopes and valleybottoms. Later in the spring, they ate new green grass and the flowers and leaves of succulent plants like dandelions. Later still, berries ripened along the river breaks and in burned areas, and grizzlies settled in for the richest time of year. As autumn waned, fresh diggings along valley slopes showed that the bears were feeding again on roots. Where there were reliable spawning runs of salmon, suckers, or other fish, the grizzlies gathered annually to feast on the seasonal bounty. And when winter's first snows swept grey across the land, the bears retreated to their winter dens on lee slopes where snowdrifts were sure to form, insulating them from the storms and cold of winter.

Grizzlies still follow their seasonal rounds in the mountains of Alberta and British Columbia, but most of the West's grizzly country lacks bears now. By the early 1900s, grizzlies had been exterminated from the prairie and parkland regions. A small population survived in Alberta's Swan Hills, but for the most part grizzlies lingered on only in the high country in the mountains of Alberta and British Columbia.

In the 1970s and 1980s, grizzly populations in the southern Canadian

Rockies and parts of southern British Columbia began to recover from earlier losses. The recovery was helped by conservative hunting seasons and the national parks introduced programs to keep bears away from garbage and other artificial food sources, and to educate park visitors about bears. Nonetheless, by the late 1980s the increasing industrialization of the West's surviving wild country was again putting stress on the wilderness-dependent grizzly.

A three-year study of grizzly bears in Yoho and Kootenay national parks nearly failed for lack of data, because radio-collared bears did not live long enough. More than half the study

A grizzly who attacks a human generally dooms itself to death ... and the population suffers a major setback, especially if that bear was a valuable breeding female.

bears died in two years. Since grizzlies rarely breed before the age of six or seven years, and then have at best three cubs every three years, this rate of mortality is a guarantee of extinction. In fact, several biologists predict that the grizzly will not survive the next century in most of British Columbia and Alberta.

Grizzly bears simply cannot afford too much mortality. Until now, the

great bear has never had to worry about predators during its entire history as a species, and so it has one of the lowest birth rates of North American mammals.

Logging, mining, and livestock production, unfortunately, bring both temptation and roads to bear country. The temptation is in the form of garbage dumps, remote camps, and easily killed cows and sheep. Since a grizzly bear's whole life is a race to accumulate fat for its winter sleep, any source of easy food is hard to resist. Once addicted to human food, a bear is doomed; eventually its addiction will put it in front of a bullet.

Industrial roads open up grizzly country to legal hunting and, worse, to poaching. Provincial governments have recently legalized the sale of wild animal parts. Bears – grizzly and black alike – are now being poached as never before. Their claws and gall bladders are as valuable as narcotic drugs, and they inspire the same kind of interest from the criminal world.

Tourism may offer some hope for the grizzly, but it also presents major threats. Tourism development brings roads and temptation just as other forms of industrial use do. In addition, hikers and others who use grizzly habitat for recreation, occasionally run into bears at close range. If a hiker encounters a female grizzly with cubs, or a grizzly guarding a food source, there is a good possibility that the grizzly will attack. A grizzly who attacks a human generally dooms itself to death ... and the population suffers a major setback, especially if that bear was a valuable breeding female.

On the other hand, tourists value grizzlies and the chance to see them, more than many other animals. If tourism develops in ways that protect grizzly habitats, and if humans can learn to use grizzly country with humility and respect, the economic value of tourism may become an argument for protecting grizzlies from more destructive forms of development.

History has not been kind to the great bear. Its future looks grim. If the grizzly survives, it will only be because we decide that it should survive.

Black Bear

THE BLACK BEAR has lived for too long in the shadow of the grizzly. Lacking the mystique of its wilderness cousin, the black bear has been called "the clown of the woods" and "vermin" – titles that demean a unique and vulnerable northern animal.

Black bears are more adaptable than grizzlies, readily coexisting with human beings. Occasionally, black bears even turn up with the city limits of Vancouver, Calgary, Edmonton, and Winnipeg. Since black bears breed more prolifically than grizzlies and occupy smaller home ranges, they are also more resilient to hunting pressure and other population stresses.

Unlike the grizzly, black bears reach breeding maturity at three or four years of age, and a female may then produce two or three cubs every second year for the rest of her life. Under ideal circumstances, a female black bear may have 21 descendants by the time she is ten years old, whereas a female grizzly may have only six at best. So, black bears can cope with a higher death rate.

But the world is changing, and black bears face a dangerous future. Historically, black bears were exposed only to legal hunting and being killed as nuisances when they developed an appetite for garbage or agricultural crops. Today hunting is well-regulated and responsible garbage management has reduced the chances for bears to become nuisances. Jasper National Park has enclosed its landfill with an electrified fence to keep bears out. Most parks and recreational areas now use bear-proof garbage containers. In some northern areas, beekeepers use electric fences to keep bears away from their hives. In fact, the future was looking good for the black bear in western Canada, until the poaching market was opened up in the late 1980s.

Ominously, some provincial government have legalized the sale of animal parts in recent years. Black bear gall bladders and claws are used as medicines and aphrodisiacs in the Orient. They can fetch hundreds of dollars on the international market. The temptation, for international poachers and for people living in depressed rural economies, is hard to resist. Bears are readily attracted to bait, and there are many abandoned roads in western Canada. For the price of a bullet, any poacher can make a big profit from a dead bear. And unlike legal hunting, there are no closed seasons or bag limits for poachers.

Black bears live in forested areas,

especially where there is a mix of conifers and deciduous trees. Landscapes that were burned twenty or more years ago are particularly valuable because of the dense cover of young trees and the abundance of rotting fallen timber. Ants are an important, energy-rich food for bears, and decomposing deadfalls provide an abundant source. For the black bear, it is the abundance of fallen timber that make old burns far better habitat than old clearcuts.

For the black bear, it is the abundance of fallen timber that make old burns far better habitat than old clearcuts.

In coastal areas where winter is wet rather than snowy, black bears may be active virtually the whole year round. However, farther inland most black bears sleep the winter away in dens under fallen trees or dug into hillsides. In the spring, having used up to a quarter of their body weight

simply to stay alive during the winter sleep, black bears emerge from the den and begin a nonstop search for high-quality food. Green vegetation is very important to them, as are ants, grubs, spawning fish, and carrion. Some large black bears become efficient predators, specializing in killing newly born calves of elk, moose, and other large animals. For the most part, however, bears prefer not to waste energy on hunting since their roly-poly design and large size makes them inefficient predators.

Black bears are generally tolerant of human beings, often feeding unconcernedly on roadside dandelions as delighted tourists jostle for photographs. Even so, the black bear is dangerous and there have been several attacks on humans in recent years. Black bear attacks tend to involve deliberate predation attempts, unlike grizzly attacks which are usually defensive. The smaller bear is, potentially, the more dangerous of the two species. All bears should be enjoyed with respect, from a distance.

Cougar

padding silently among the shadows, stalking and killing deer and elk, watching from the trees ... the cougar adds a unique, wild flavour to the western landscape.

Cougars are far from rare in western Canada, although they have vanished from most of their prairie range and the eastern race of the cougar may be extinct. The cougar thrives in the foothills of western Alberta and in the dry interior valleys and the temperate coastal rain forests of British Columbia. Since it is largely nocturnal, the cougar is rarely seen.

Cougars are large animals, with some males reaching lengths of 2.4 meters and weights approaching 100 kilograms. Like most cats, they hunt by stalking and can bring down prey as large as moose. In fact, in western Alberta moose can comprise up to half the winter diet of some male cougars. However, cougars prefer to eat deer. The highest densities of cougar coincide with the winter ranges of mule deer in Alberta and British Columbia. Beavers, porcupines, and snowshoe hares can also be part of a cougar's diet.

Humans also figure in the list of cougar prey items. There have been several cases of cougars attacking and killing children on Vancouver Island and in other areas. Although the number of cougar attacks is very low, it will be interesting to see if the growing trend for affluent families to build homes on subdivisions in high-quality deer habitat leads to more attacks. When deer hunting is banned in rural

I NEVER REALLY believed in cougars until one June day in Kootenay National Park. I was in the bunkhouse packing for a field trip when I heard a high-pitched yapping and howling outside. Stepping to the door, I looked out. A small coyote, every hair on end and its tail swollen up like a bottle brush, was standing just off the front step, looking up the hill at the forest, yapping in a strange, high-pitched way.

Jim Mulchinock joined me, and we opened the door and stepped out. Less than twenty yards away, a huge tawny cat was sliding silently out of the forest shadow, intent on the coyote. At the sound of the door opening, the cougar stopped, fixed Jim and me with a long stare, then turned and disappeared back into the undergrowth.

The forests and mountains of the West have never really looked the same to me since then. There is something thrilling about the realization that those giant wild cats are out there, residential areas, deer populations often increase markedly. This, in turn, supports higher densities of cougars.

If people and cougars are to continue to coexist in the Canadian West as successfully as they do today, it will be necessary to identify and protect important deer and other wildlife habitat through zoning restrictions and other kinds of land-use control.

Lynx

IT MAY NOT TURN white in winter, as its favourite prey does, but few animals are as specialized for life in the northern winter as the lynx.

Lynx tracks in soft snow are as large as, or larger than, cougar tracks. The average lynx, however, is less than a quarter the weight of a cougar. The reason for the large tracks is simply that the lynx has huge feet that function as snowshoes, allowing this tufted-eared cat to hunt in deep snow country without wasting valuable energy breaking trail.

In fact, lynx tracks are rarely seen because these wild cats travel on the packed trails of snowshoe hares much of the time. Few northern predators can afford to be specialists, but the lynx clearly specializes in hunting and killing snowshoe hares. As a result, lynx populations rise and fall in response to the cycles of abundance that typify hare populations. One year lynx seem to be everywhere; the following year they seem to be extinct. For several years their numbers may increase. Until once again, reflecting snowshoe hare populations, the lynx population crashes.

Rarely, lynx become exceedingly common at the peak of their population cycle. One winter in the late 1960s, more than 60 different lynx were captured or shot within the Calgary city limits. Of the five lynx I have spotted in my life, three were seen that year.

Besides hares, lynx also eat grouse, squirrels, mice, and other animals when they can catch them.

Since hare are commonest in young forests and dense spruce and willow thickets along river bottoms, these are also the most important habitats for lynx. Look for lynx in areas that were burned from 10 to 30 years previously and in the productive vegetation complexes along northern rivers. Conservation of lynx means protecting valley bottoms, particularly north of the Yellowhead Highway, and allowing wild fires to occasionally burn across the northern forests.

The lynx is a northern animal that occupies the broad band of coniferous and mixed forest that covers most of northern Manitoba, Saskatchewan, Alberta, and British Columbia and extends south through the mountains to the United States.

Bobcat

BOBCATS ARE rare in Canada, largely because they do poorly in deep snow country. For the most part, their range coincides with the driest parts of the West: the southern prairies and the interior valleys of British Columbia.

Bobcats are generally about two-thirds the size of the similar-looking lynx, but their feet are much smaller and they have difficulty travelling in deep, soft snow. Like the lynx, however, they concentrate their hunting efforts on rabbits, feeding on snowshoe hares and cottontail rabbits. A recent study in southern Alberta revealed the surprising fact that bobcats are responsible for up to half the mortality of pronghorn fawns. At a certain stage in the pronghorn's development, fawns have not yet learned to run from predators but have abandoned the habit of lying flat to the ground. Bobcats cue in on these vulnerable fawns and hunt them effectively. In some cases, the little wildcats have even been known to kill adult deer, but for the most part they eat rabbits, mice, birds, and other small animals.

Bobcats are mostly nocturnal, but they can occasionally be flushed from their day-beds in the dense shrubbery along prairie coulees and rivers, or in rocky slopes in southern British Columbia. Police Coulee, in Alberta's Writing-on-Stone Provincial Park, as well as the streams that drain from the Cypress Hills in Alberta and Saskatchewan, contain populations of bobcats.

Like lynx, bobcats generally have a single litter of kittens each year. Starvation is the biggest threat to bobcats as they grow up; the first winter is particularly crucial since the kittens have not yet reached their full adult size and consequently need more energy while having less talent at capturing food. Northern winters can decimate bobcat populations since this cat is poorly adapted to snow. Global warming may have the effect of increasing the success of western Canada's bobcats at the expense of the lynx and other deep-snow animals.

Coyote

LATE AT NIGHT, almost anywhere in western Canada, the song of the coyote can be heard. High-pitched yapping and maniacal shrieking make it sound like a wild party is going on somewhere out there in the darkness. Then silence falls, and wide-eyed children feel about in the blackness of their tents to make sure their parents are close by.

Humans have little to worry about from coyotes, although there have been a number of attacks in recent years, mostly on small children. Mice, voles, hares, birds, grasshoppers, and other larger animals including deer, have cause to listen intently to those midnight howlings. The coyote is an effective and highly adaptable northern predator.

Coyotes rarely establish packs to the same degree that their larger relative the timber wolf does. However, pairs of coyotes can still execute well-planned hunts. I have watched coyotes take turns chasing jackrabbits and team up to try and separate bighorn lambs from the herds. More often, coyotes hunt opportunistically and alone, catching whatever is available.

In Jasper National Park and other mountain areas, however, packs of coyotes are not an uncommon sight in winter. One winter a heavy snowfall ended with a brief thaw, and it was followed by an extended bitterly cold spell. The crusted snow was too weak to support the weight of elk, deer, or wolves, but the smaller coyotes had little trouble with it. During that week coyotes killed several deer.

I witnessed one of the kills. Four coyotes surprised three doe mule deer from above. When the coyotes erupted from the trees, two of the deer stood their ground, lowered their heads and faced the charging coyotes. The third deer turned and fled. It was a fatal mistake, because it allowed one of the coyotes to bite at her flanks several times before she plunged into the Athabasca River.

The coyotes waited until the injured doe clambered out on the far shore before following her across. They cornered her in a stand of pine trees. Whenever one of the coyotes darted close to her, however, she jumped at it and kicked with her forefeet. The coyotes would then retreat a little ways and lay down. The doe stood quietly, head down.

Eventually she tried to lie down. Immediately, the coyotes closed in on her, and again she was forced to fight them off.

However, time was on the coyotes' side, and the doe's wounds weakened her. At length she fell to the assault, and the coyotes fed. Later, ravens, magpies, and even chickadees gathered to pick over the remains.

Coyotes rarely find it so easy to kill large animals, and more often they scavenge from the kills of wolves, cougars, and other large predators.

Coyotes breed in midwinter, and the female gives birth to around six pups the following spring. Although the pups can be eaten by eagles, wolves, cougars, and lynx, by far the greatest threat to the growing pups is

Due to their adaptable nature and innate intelligence, coyotes have little to worry about in the long run.

starvation. Severe winters, droughts, or periodic crashes in the numbers of hares or voles, can set coyote populations back temporarily. Due to their adaptable nature and innate intelligence, coyotes have little to worry about in the long run. Coyotes are one of few native animals in western Canada that have thrived in spite of changes humans have wrought over the past century.

Wolf

I WAS COUNTING migrating birds along a survey route that followed the Athabasca River north from the town of Jasper when I stopped beside the busy Yellowhead Highway to scan the floodplain of the Rocky River.

Like many western rivers, the Rocky traces a braided pattern across open gravelly flats studded with driftwood tangles. The river is too small and volatile to carry all the rocks and debris that wash down from the high mountains, so its valley is gradually filling with gravel.

Out on the flats I spotted something black. I set up the spotting scope for a better look. It was a wolf, staring at me from behind a clump of weathered driftwood a couple hundred metres upstream. Another wolf raised its head, then moved out into the open.

The wolves were eating something behind the driftwood, and for several minutes I watched them, delighted. Suddenly, one took a few steps out onto the gravel flats and stared, alertly, at the far bank.

A grizzly and her two cubs had emerged from the timber and were heading across the wind-whipped flats.

The wolves retreated before the bears, but they didn't leave. As the sow and her cubs settled themselves for a short feed, the wolves lay down and watched from a few metres away.

This was too good to keep to myself; I decided to flag down someone to share it with. The first car to stop contained a woman from Toronto and her teenaged daughter. They could hardly contain their excitement as they watched the grizzly and her cubs wander back into the forest, and the wolves return to their kill.

After they had left, another vehicle stopped. This one contained a retired couple from northern Saskatchewan. The woman watched the wolves briefly, then stood aside while her husband took his turn.

"I don't see any grizzly bears," he said.

"They left a few minutes ago."

"Well," he snorted, "you mean you stopped us to look at some goddam wolves? I spent half my life trapping those things; I'd be just as glad if I never see another wolf again."

The timber wolf brings out extremes in humans. Few people are neutral about wolves; either you love them or you hate them. To a large degree, our feelings are shaped by our experiences. Those who raise livestock in wolf country, or who compete with wolves for elk and deer each hunting season, are unlikely to think kindly of them. Those whose lives are more remote from nature are often more inclined to idealize the wolf as one of the most noble and romantic of wilderness animals. "Only the mountain," wrote the great American conservationist, Aldo Leopold, "has lived long enough to listen objectively to the howl of a wolf."

In the 1950s and early 1960s, in response to a rabies outbreak in foxes and skunks, intense poisoning campaigns were carried out across western Canada. Even the national parks were not spared; some park wardens were reprimanded for not killing enough predators! Already exterminated from the prairies and river valleys of southern Manitoba, Saskatchewan, and Alberta, wolves vanished from much of their remaining west-

... killing wolves is more expensive ... than simply using wildlife habitat in a responsible, ethical manner.

ern range south of the Yellowhead Highway.

However, since 1975 wolves have made a tenuous recovery over much of their range. In fact, Canadian wolves have even recolonized northern Montana. Visitors to Kootenay, Banff, and Jasper national parks occasionally spot wolves hunting in the valley bottoms, or they hear them howling at night. On Vancouver Island, too, the wolf has recovered from near-extinction.

Wolves hunt deer, elk, and moose as well as smaller prey. In some areas, wolf predation combines with habitat loss, human hunting, roadkills, and other factors to keep ungulate populations at low levels. This is what has happened to caribou in central Alberta.

On the other hand, a healthy prey population in good habitat can readily recover from wolf predation, human hunting, or the occasional severe winter. The key is good habitat and healthy animals. Killing wolves to build up deer, elk, and moose populations is easier, politically, than changing the way humans use wildlife habitat. However, killing wolves is more expensive, and less effective over the long term, than simply using wildlife habitat in a responsible, ethical manner.

A wolf pack typically contains a dominant male and female who produce a litter of from two to ten pups each year. The other pack members help feed the pups. Whether the pups survive to maturity or not depends on how abundant their prey is, how severe their first winter is, and how vulnerable the pack is to poisoning, legal

hunting, mange, etc. Wolves can cope with most forms of persecution, other than poisoning.

Wolf packs live in well-defined home ranges, some as large as a thousand square kilometres. Individual wolves may range much farther; a young female wolf from northern Montana was recently shot near Fort St. John, British Columbia. This wide-ranging behaviour helps wolves re-colonize areas where they have been wiped out. But even so, it is unlikely that viable numbers of wolves will ever again inhabit the southern interior of British Columbia or the prairie and parkland regions of Alberta, Saskatchewan, and Manitoba.

Red Fox

niche formerly occupied by the swift fox. A recent program by the Canadian Wildlife Service to reintroduce swift foxes to the last few remnants of prairie wilderness has been complicated by the fact that coyotes and red foxes kill them.

Farther north, it appears that the red fox may also suffer from persecution by its larger relatives. There are few areas where wolves, coyotes, and red foxes coexist. Red foxes in the northern wilds often depend on wolf kills for winter food, scavenging on

THE RED FOX is found throughout the Canadian West, but it is rare in the higher mountains. The first one I ever saw was sitting at the mouth of its den in an overgrazed horse pasture beside the Trans-Canada Highway east of Calgary. In fact, the adaptable red fox is just as much at home on the outskirts of a foothills town or in a hole dug beneath a prairie granary as in the forested wilds of northern Manitoba.

All the fox asks of life is an abundant supply of small things to eat: grasshoppers, mice, fruit, and rabbits. Like most carnivores adapted to life in northern climates, the red fox is an opportunist. It eats whatever the season, and fate, provides.

Western Canada is home to four native members of the dog family – the wolf, coyote, red fox, and swift fox. Swift foxes were exterminated from the prairies early in this century, largely as incidental victims of poisoning campaigns aimed at coyotes and wolves. Red foxes inherited the enviromental the remains after wolves are finished. Coyotes do the same thing, and where coyotes and wolves coexist, foxes are scarce.

Until the settlement of western Canada, foxes were mostly confined to the northern forests. When British remittance men arrived in the western plains during the late 1800s, they brought a taste for fox-hunting over hounds with them. Unable to find any red foxes, however, they had to make do with coyotes.

In Jasper National Park, wolves, foxes, and coyotes can all be seen along the Athabasca Valley, especially north of the town of Jasper in winter. Even so, the foxes and coyotes appear to have segregated; foxes appear to dominate the Jasper Lake area, while coyotes are most common closer to town.

Red foxes sometimes carry rabies, and frequently are afflicted with mange. Any seemingly friendly fox should be avoided and reported to wildlife authorities.

Wolverine

ONE OF THE highest densities of wolverines in Canada is in the timberline country along the spine of the Canadian Rockies. Visitors to Banff National Park, jockeying for a parking stall at Lake Louise or Moraine Lake, might be excused for believing that wilderness is more myth than reality, but they are in the heart of wolverine country and the wolverine, perhaps more than any other animal, is truly a wilderness creature.

Wolverines are reputed to be vicious, but they aren't. What they are is determined. They are mid-sized carnivores that spend each winter in deep-snow country long after most wolves, cougars, and other carnivores have departed for winter ranges where the hunting is easier. The wolverine's entire life is a serious, no-holds-barred quest for food. Even so, wolverines are virtually never aggressive to humans.

In Kootenay National Park, Larry Halverson followed the sound of repeated crashes near Floe Lake and discovered a wolverine throwing itself bodily against the locked door and shuttered windows of a warden patrol cabin. When a wolverine succeeds in getting into a building it consumes anything edible it can find and generally leaves the shambles stinking of musk; like skunks and other weasels, wolverines have foul-smelling glands at the base of their tails that they use for marking anything they consider to be their property.

If a wolverine finds the carcass of a large animal killed by an avalanche or some other predator, it will gorge itself on all it can hold, and then carry off bits of the carcass and hide them for later use. Wolverines, perhaps more than other members of the weasel family, scavenge from carcasses at any chance they get; life in the snowy northern wilds is not for the fastidious. In some cases, wolverines have been seen to bluff wolves away from animals the wolves have killed. In Banff, however, wolves have also been recorded as having killed wolverines that came too close. A Canadian Wildlife Service crew in northern Banff National Park once watched a wolverine circling a dead elk that was being fed upon alternately by a family of grizzlies and a pack of wolves; whenever the larger animals were on the kill the wolverine stayed back.

In summer, wolverines enjoy an easier life, feeding on berries, eggs, small animals, and insects. The long circuitous journeys of winter are replaced with shorter trips, especially for breeding females who have a litter of two to four cubs in late winter.

Wolverines are survivors, living in remote terrain and deep snow where food is often scarce and conditions extreme. They are not highly productive, and they have high natural mortality. Consequently, trapping can deplete wolverine populations rapidly, and this is exactly what has happened over most of the animal's natural range over the past century and a half.

Wolverines are reputed to be vicious, but they aren't. What they are is determined.

Wolverines are rare in Canada's northern forests and have recently been declared endangered in the eastern provinces. Under the protection of the mountain national parks, however, healthy populations of this wilderness animal still survive.

American Marten

THE OLD-GROWTH forests of the western mountains and northern prairie provinces are home to the marten, and its less-common larger relative, the fisher. Unfortunately, old-growth forests are becoming increasingly rare as logging companies replace them with young stands of fast-growing commercial trees. As a result, the future of the American marten looks likely to be restricted to the protected forests of national parks, unless public pressure forces logging companies to protect marten habitat.

If it were not for the fact that traditional forest management practice is to replace old forests with young trees that will never again be allowed to grow old, the future of the marten would actually be more hopeful than it has been for the past century. The marten is known for its excellent fur, which is very similar to the European sable. It was a mainstay of the trapping industry for years, since its fur is light, easy to prepare, and valuable.

Recently, however, changing tastes in Europe and the United States have resulted in a precipitous decline in fur values. Trapping is no longer economic, freeing martens from this source of mortality.

However, nothing is simple. Much of the most effective pressure on logging companies to spare old-growth timber and to protect wildlife habitat has traditionally come from trappers. Trappers, in their annual forays into the backwoods, were often the first to raise the alarm when new logging roads and clearcuts invaded marten habitat. Now that trappers are forced to rely on wage labour or unemployment insurance, who will watch over the interests of the marten and the other fur-bearers?

Martens vary from reddish-brown to a rich chocolate colour, usually with a pale marking on their chest. They are curious animals, and will often approach for a closer look in response to squeaking noises, even when they can clearly see that the noises come from a human. In winter, their paired tracks wind everywhere through the snowy winter woods, tracing the course of their endless hunt for voles, mice, hares, and squirrels.

A female marten usually will produce a family each year, but the average brood size is only three young. High-quality habitat – old-growth spruce and fir forest with scattered small openings – should be jealously guarded by people who value the chance to encounter this curious and handsome mammal.

Weasels

THE DAY I MET my wife, we watched a short-tailed weasel come within a few centimetres of killing a warbler beside the outflow from the Cave and Basin hotsprings in Banff National Park. It was a sunny day in late May. A late storm had surprised the landscape, plastering newly sprouted foliage with heavy, wet snow, and forcing migrating birds to forage on the ground. In a grassy clearing beside a small hotspring, a Wilson's warbler was hopping unconcernedly about when the weasel emerged from the bushes.

The slender brown animal appeared unaware of the warbler. It bounded about the clearing, weaving and leaping, looking for all the world as if spring fever had put all thought of predation out of its mind. Sometimes its jumps brought it close to the bird, other times it leaped directly away from it. Suddenly, it leaped sideways and snapped at the warbler so fast it took my mind a moment to catch up with what my eyes were seeing.

Fortunately for the warbler, its reflexes were better than mine. It flitted into the bushes. The weasel, acting as if nothing had happened, went leaping away into the undergrowth and was gone. I was left with the impression that a weasel does not distinguish between playing and killing.

There are three kinds of weasels in western Canada. The least weasel is the size of a mouse. Its tail is short and lacks the black tip that distinguishes the other two. Least weasels are found mostly in bogs, muskegs, and wetlands across the forested north.

The commonest weasel is the short-tailed. About the size of a squirrel, the short-tailed weasel is also commonly called the ermine in winter, when it is pure white with a black-tipped tail. In summer it is brown with a white belly, but retains the black tip on its tail. Short-tailed weasels prefer brushy areas along the edges of forests and wetlands.

The largest of the three weasels is the long-tailed weasel. It looks somewhat like the short-tailed but has a buffy yellow belly in summer and prefers grassier habitats. Long-tailed weasels are common in the prairie and parkland regions of the prairie provinces, but less common in the mountains than the short-tailed weasel.

Weasels are dedicated predators, feeding on whatever they can catch. The bulk of their prey consists of voles and mice. Long-tailed weasels also eat ground squirrels. An adaption for hunting and killing is so much a part of these little animals that they will kill far more than they need to eat. Back when prairie farms often included chicken coops, a long-tailed weasel could easily wipe out several dozen chickens in one night's frenzy.

Back when prairie farms often included chicken coops, a long-tailed weasel could easily wipe out several dozen chickens in one night's frenzy.

Weasels are rarely seen because they do most of their hunting by night. In winter, not only does their white coloring make them even harder to spot, but they spend a lot of time under the snowpack, hunting voles and mice in their runways between the snow and the frozen ground. A few flecks of blood at the entrance to a mouse hole are often the only hint that weasels are present.

At other times, especially when the snow is shallow and the weather is fairly mild, weasel tracks may seem to cover the entire countryside. One or two weasels, in the course of a busy night's hunting, can make enough tracks to fool the casual skier or hiker into thinking that these little carnivores are everywhere.

Winter tracks of weasels are distinctive. Their trails wander across the snowy landscape, from brush pile to tree to boulder, each pair of footprints staggered with one imprint slightly ahead of the other. Weasels bound lithely across the snow, their hind feet fitting almost perfectly into the marks of their front feet. The other similar-sized tracks in the winter forest – those of the red squirrel – generally show two small forefoot tracks side by side, just slightly behind two larger hind foot tracks.

Since weasels are small, constantly active, and do not hibernate, the obsession with hunting helps them meet their high energy needs. They are beautiful, fluid animals and an important part of the ecosystems they occupy. Starvation, owls, other predators, and roadkill all take a heavy toll on weasel populations; consequently it is fortunate that these small predators produce up to nine offspring each year.

American Badger

WHERE GROUND squirrels abound, badgers usually can be found too. In western Canada, this means that badgers are found primarily on the prairies, on open parklands, in low valleys in the Rocky Mountains, and in the dry parts of interior British Columbia.

The badger is a member of the weasel family, but its lifestyle is uniquely its own. Unlike the wide-ranging marten, wolverine, and weasel, the badger occupies a smaller home range. They do not hunt their food above ground; instead, they dig for it. Badgers are among the most impressive digging animals in Canada. Their bodies are flattened, their legs are short and powerful, and they have long, curved front claws for digging, and short, spade-like hind claws for scooping. When a badger excavates a gopher or ground squirrel from its underground burrow, it plugs all the entrances but one, and then it churns its way into the earth. Re-searchers have found that badgers are successful three out of every four attempts, a much higher success rate than most predators. In fact, the opportunistic coyote has learned to capitalize on the badger's effectiveness by simply tagging along. On several occasions along the Trans-Canada Highway near Medicine Hat, I have seen coyotes following badgers hopefully, waiting for the chance to pounce on an excavated ground squirrel.

Although badgers are adapted for digging out small rodents, they eat grasshoppers, bird eggs, and anything else they can find above ground.

Their short legs may be well-suited for digging, but they are not much use for running. A badger surprised more than a few yards from the nearest burrow is easily cornered. Once trapped, it puts on a brave display, backing up and baring its fangs at its tormentor, hissing the whole while. The display is not all bluff; many a domestic dog has come limping home the loser in an encounter with a badger. Few animals care to press the attack against a hissing badger. Consequently, for the three or four babies born to a female badger each year, starvation is a greater worry than predation.

Humans complicate badger life, because badgers are easy targets for vandals with rifles and because the roads we build all over the landscape are deathtraps for the slow-moving, nocturnal badgers. The best thing we could do for the badger would be to build fewer roads.

Striped Skunk

WHILE FEW PEOPLE ever see the nocturnal skunk, many have smelled it. Skunks like the tall vegetation and wet places often found along roadsides. Late at night when these black and white weasels are snuffling about looking for food they often fall victim to passing cars. The characteristic musky smell of a flattened skunk often lingers for days, a reminder to all passers that the pavement humans love so well spells death to most other animals.

Skunks are, fortunately, fairly prolific. A female skunk may begin breeding as a one year old and average six to eight babies each year. The wetland habitats they occupy are kept from being overrun by skunks only by the fact that in addition to automobiles, owls, coyotes, and foxes prey on these small carnivores. Nonetheless, population explosions of skunks occasionally occur, and the dispersing young invade towns and cities. High populations often coincide with high levels of disease (just as they do in humans), with leptospirosis and rabies being among the commonest diseases in skunks. It appears that skunk populations are one of the reservoirs of rabies that keeps this disease from disappearing in western Canada.

Skunks eat a variety of foods, being more omnivorous and opportunistic than most other weasels. They are important predators on the eggs of ducks and other waterfowl. But they also eat insects, earthworms, clams, salamanders, frogs, small rodents,

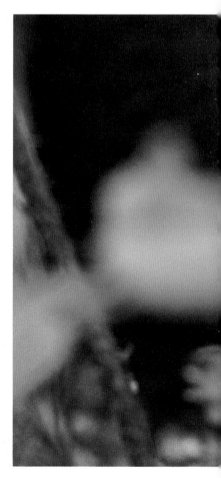

berries, and grasses. Unlike most other members of the weasel family, western Canada's skunks spend much of the winter asleep in underground dens rather than ranging afield in search of food.

Skunks are neither fast on their feet, nor equipped for attack. However, two little scent glands at the base of their tail more than make up for their lack of other defenses. A skunk surprised by a human or other large animal will usually try to simply ignore the other animal. However, if pressed, the skunk will stamp its feet, click its teeth, and growl or hiss to warn the other animal away. If that fails, the skunk twists sideways, so that both its head and its tail point at the enemy, and sprays two jets of foul-smelling musk with uncanny accuracy. The musk stings exposed membranes and takes weeks to wear away. One encounter is all it takes to educate most creatures.

Mink

IN ALBERTA's Kananaskis Country one day, I was driving along a backroad when I crossed a wooden bridge over a small trout stream. Casually glancing into the pool below the bridge, I was startled to see a large splash. Since I was going trout fishing, I stopped abruptly; any fish that could disturb that much water was worth investigating!

Backing up, I parked on the bridge and peered into the water. Something large and dark was swirling around the pool with incredible grace and speed. There was another splash, and a dark brown, dripping-wet animal squirted up onto the bank, ran a few steps, and dove back in.

The mink moved with an underwater grace I had never imagined, probing and hunting among the boulders at the bottom of the pool. Finally, he flushed a small cutthroat trout and chased it the length of the pool and back again before it escaped. A moment later, the mink emerged onto the bank and disappeared into the undergrowth, only to reappear a few moments later, farther downstream.

Mink are widespread throughout western Canada, from the sedge-lined sloughs of the Saskatchewan prairies to the subalpine stream banks of Jasper's Tonquin Valley. They thrive anywhere that land and water meet, especially if it is a gentle meeting involving quiet backwaters, sedge meadows, cattails, or mudflats. Mink eat frogs, fish, insects, voles, small birds, and crustaceans. Their hunting constantly takes them from dry land to underwater and back again. They are as at home along the Pacific coast as they are in inland habitats.

Solitary hunters, mink travel almost constantly and mostly at night. Stream-dwelling mink may have home ranges that extend along two to five kilometres of stream. After a brief mating season in midwinter, the females give birth to up to eight babies in April and raise them alone. Late in July, the young mink set off on their own into a world where their chief enemy appears to be humankind. Mink are vulnerable to pollution-related illnesses, since they feed on aquatic animals that concentrate water-borne pollutants. They are also vulnerable to trapping. But trapping pressure has decreased over the past few decades due to the animal rights movement, and the rise of the mink-ranching industry, which has come to dominate the market.

River Otter

IRONICALLY, ONE of the best places to see the river otter is not along a river but at the seashore. River otters eat fish, and nowhere is fish more readily available than the estuaries and shallow bays of the Pacific coast, especially where streams empty into bays. At low tide, river otters can often be spotted at Grice Bay in Pacific Rim National Park.

A related species, the sea otter, was nearly exterminated because of its valuable fur. Small populations have been re-introduced along the west coast, where they live offshore among the giant kelp beds.

Inland, the river otter is widespread, but rarely seen. Prairie rivers, however, hold only the ghosts of otters past; the species was extirpated from most of the Saskatchewan and Missouri river watersheds during the late 1800s. The otter has always been rare in the high mountains, simply because there is little suitable habit.

But it is still fairly common along the great northern rivers such as the Peace, Athabasca, Churchill, and Nelson, and many smaller northern streams.

Otters are incredible swimmers, chasing down and killing fish underwater. They prefer slower fish like suckers, spawned-out salmon, carp, and squawfish to faster species like trout and pike. As well otters feed on waterfowl, eggs, and small animals that live along stream floodplains.

Otters are noted for their playfulness, sometimes sliding repeatedly down mud banks like kids on a toboggan run, or alternately running and sliding in snow as they travel up and down partially frozen winter rivers. At this season their tracks are distinctive, with the typically staggered gait of most members of the weasel family, and with frequent marks of belly-drag.

Otters give birth to around six babies late in the winter, generally in a den dug by some other animal, or in a natural shelter like a log jam or abandoned beaver lodge. Two to three months later, the young otters begin to wander farther afield and replace their mother's milk with solid food. Eventually, the whole family moves on, following the typical otter lifestyle of nighttime explorations along their territorial river or stretch of seashore.

Northern Sea Lion

ANIMALS THAT LIVE mostly in the water are often highly specialized to an aquatic lifestyle, to the point where they look ridiculous on dry land. The sea lion is a classic case; incredibly graceful and playful in the water, sea lions are painfully slow and awkward on land.

Sea lions are so well-adapted to life at sea that in stormy weather, when we humans pull our boats well up onto land, tie them down, and retreat to our dry houses, sea lions abandon their rocky islands and headlands for the safety of the swells. Sea lions gather in large herds on their rocky refuges during sunny afternoons, but spend most of their time in the water.

A male sea lion may weigh more than a tonne and be more than three metres long. Dominant males gather harems of females during the sum-mer breeding season, while younger males keep a safe distance. The price that the dominant males must pay for their breeding privileges is starvation. A dominant male sea lion may go for two months on the breeding ground without eating anything. In Canada, most northern sea lion rookeries, as the breeding grounds are known, are in the Queen Charlotte Islands and other isolated islands along the north-ern coast of British Columbia.

Female sea lions give birth to a single pup more than twelve months after breeding. The babies mature rap-idly, being able to swim and hunt a few days after birth.

Sea lions feed on a wide variety of sea life but, unlike the smaller seals, they eat few salmon or other com-mercially important fish. Even so, they are looked on with suspicion by many commercial fishermen and for most of this century sea lion populations have been controlled by killing to re-duce their perceived impact on fish populations.

Other than humans, the only sig-nificant enemies that sea lions have to contend with are killer whales and some of the larger shark species.

Sea lions carry a thick layer of fat immediately beneath their fur. And the fur grows in dense, oily clumps. The combination insulates the warm-blooded sea lions from the cold north Pacific Ocean.

Harbour Seal

HARBOUR SEALS travel long distances up British Columbia rivers each spring in pursuit of spawning salmon. One population lives permanently in Harrison Lake, almost 100 kilometers from salt water. In the days before the great dams blocked the Columbia River and killed many of the river's salmon runs, seals may have travelled much farther inland.

These small seals are about 1.5 metres long and weigh 50 to 75 kilograms. The similar-sized fur seal lives far out to sea, rather than close to shore like the harbour seal and the sea lion.

Harbour seals are common along the British Columbia coast where they are often seen in groups resting on sand bars, rock outcrops and jetties, or bobbing about in the surf watching the world go by. Unlike the larger sea lions, male harbour seals do not gather and defend harems of females. Instead, the seals mate in a more haphazard manner during late summer, with the females giving birth to one, or sometimes two, pups almost a year later.

Baby seals spend most of their time in the water, perhaps because they are vulnerable to eagles and other predators when out of the water. However, adult harbour seals have few enemies; the most important threat are pods of killer whales that hunt the near-shore waters and are very effective as seal predators.

Human beings are also enemies of the harbour seal. Since seals feed almost exclusively on fish, commercial fishermen and government fisheries managers feel little love for them. Seals are adapted to take advantage of the same concentrations of fish that fishermen exploit. In fact, the adaptable harbour seal readily takes advantage of gill nets and other fishing gear to get free meals. As a result, there are regular programs in place to keep populations of harbour seals below their potential, in order to reduce competition for fish like salmon, herring, and cod.

Harbour seals, like most animals that feed upon fish (humans included), are vulnerable to water pollution. Pesticides, industrial chemical discharges, and other contaminants inevitably find their way into streams and rivers that discharge into the sea. There, filter-feeding invertebrates concentrate the pollutants in their bodies. Fish, eating the invertebrates, build up even greater levels of contamination, mostly in their fat and organs. Seals, whales, and humans, at the top of this food chain, end up with the highest level of contamination. This process of concentrating pollutants at higher and higher levels in the food chain is called bio-accumulation and, ethical reasons aside, is one obvious reason why humans ought to have a vested interest in eliminating pollution.

Killer Whale

THE KILLER WHALE or orca is one of the most strikingly beautiful marine mammals, with its elegant pattern of black and white markings.

Pods of killer whales range the bays, estuaries, and fiords of British Columbia, hunting salmon, cod, herring, and other sea life, including sea lions and even the huge grey whale. Each pod is an extended family group led by a female who may be as much as a century old.

Despite its name, the killer whale is not considered dangerous to human beings. In fact, in the past few years, whale-watching has become a popular activity. Boatloads of excited viewers locate a pod of killer whales and then settle down to watch from a respectful distance, so as not to disturb them. Although killer whales may not look upon humans as potential prey, there are few other marine creatures that they will not attack.

Recent research suggests that the larger whale pods feed mostly on fish, but that small pods and single killer whales are more aggressive feeders, often attacking sea lions, seals, porpoises, and other whales.

Killer whales are more closely related to the porpoises than to other whales. Like dolphins, they can leap right out of the water, but, at lengths exceeding seven metres and weights of up to five tonnes, the killer whales create a much more dramatic re-entry. Merely to break clear of the water, they must build up a speed of more than 35 kilometres per hour.

A killer whale pod is a unique social unit. Pods are restricted to closely related animals which never leave their home pod to join another. Occasionally, two or more pods will join together for a while, but when they separate there is no mixing of members. Because of the rigidity of this social system, new pods are rarely established and inbreeding is common. Biologists have determined that there are about 30 killer whale pods around Vancouver Island; about half contain six or less killer whales.

One of the best-known areas for watching killer whales is Robson Bight, a narrow sea passage near the mouth of the Tsitika River, on the east side of Vancouver Island. Here, for some reason known only to the whales, killer whale pods visit "rubbing beaches" of small, clean rounded pebbles. Perhaps because they are so vulnerable in the shallow water, or for some unknown reason, the orcas are reluctant to be watched by people on the shore. Viewers in boats anchor well back from the whales and turn off their engines.

Robson Bight may not be used much longer, however. As forest companies continue to liquidate Vancouver Island's old-growth forests, they are pushing roads and clearcuts deeper into the last remote watersheds. The headwaters of the Tsitika are now being logged. This may result in the river bringing mud to foul the rubbing beaches, as raw clearcuts and roads are eroded by the frequent rains in the area. Also, more visitors will be able to reach the "rubbing beaches"

... a pod of killer whales cruising silently through the mist is, in some ways, a distillation of all that makes Canada's biologically rich West Coast unique.

by logging road and a short hike, disturbing the whales.

The sight of a pod of killer whales cruising silently through the mist is, in some ways, a distillation of all that makes Canada's biologically rich West Coast unique.

Grey Whale

ers on Long Beach at Pacific Rim National Park sometimes spot plumes of white spray far out beyond the breaker line. They mark the exhalation of a grey whale. Occasionally, lucky watchers see a whale spy-hopping (sticking its whole head out of the water) or breaching the surface.

However, not all the whales go to the sub-Arctic seas. A few grey whales remain off the west coast of Vancouver Island throughout the entire summer. There they feed on shrimp-like amphipods and other soft-bodied creatures that live on the ocean floor. Their feeding strategy is to act like a vacuum cleaner, mooching along the sea floor, churning up silt and mud as they strain their food from the sediment. Grey whales eat only during the summer.

Once threatened with extinction by the whaling industry, grey whales have been protected since 1937. There are now more than 16,000 grey whales ranging Canada's Pacific Rim. With commercial whaling banned, their major enemies now are the smaller killer whales. They attack grey whale calves and, occasionally, they kill the adults by attacking their fins and tongues.

The similar-sized humpback whale may once have been nearly as common as the grey whale, but it was more vulnerable to the whalers who operated out of many coastal communities in the 1800s and early 1900s; it has never recovered from the slaughter.

LIKE A LOT OF other northerners, the grey whales of the cold northern Pacific Ocean prefer to spend their winters in balmier climes. Late each autumn, thousands of these 10 to 15 metre leviathans migrate along the British Columbia coast on their way to their winter home in the warm lagoons off Baja California.

Unlike other migrants, however, the grey whales use their southern sojourn to give birth to their babies. In the spring, the males and sexually immature whales head north first, followed by groups of females with their single babies. Northward migrants appear off Vancouver Island in April and May. Continuing north, usually within view from the coast, the whales follow the breakup of the northern ice pack to their summer homes in the Beaufort, Bering, and Chukchi seas, far to the north.

Migrating grey whales are regularly seen off the west coast of Vancouver Island. Tourists watching surf-

Whatever people do on land, the

results usually end up in the ocean. Grey whales swim close to shore where they are close to river mouths and coastal industries. And because they feed on millions of small animals that filter the sea water, grey whales concentrate pollutants in their body fat and organs. Recent mysterious grey whale deaths appear to have been at least partly due to contamination from manmade chemicals. The continued

A few grey whales remain off the west coast of Vancouver Island throughout the entire summer.

health of Canada's grey whales will depend on our willingness to force industries, towns, and cities to eliminate pollution and to ensure that the water that reaches the ocean is as clean as it would be naturally.

Epilogue

We live in a world that is changing more rapidly than ever before. Every day, dozens of species of plants and animals beome extinct. Every day, unique natural ecosystems are destroyed or simplified beyond repair by industrial pressures.

Change is inevitable. Increasingly, change is the result of human decisions, not ecological processes. Each time we choose to buy a product, support a politician, or ignore a problem, we are influencing the rate and degree of change. In a world where all things are connected, everything we do has an impact on wild animals and the habitats they utterly depend upon. We need to ask ourselves some vital questions.

What kind of future do we want? How many of the things that enrich our lives today do we want to carry with us into the future? And what do we need to do to insure that the things we value survive?

We are only as good, and as rich, as the environment in which we dwell. We cannot be citizens of a place that we have helped to ruin. Canada's unique mosaic of natural ecosystems, with their assemblages of native animals, are what make us Canadians. The same thing is true for Americans, Britons, Japanese, Germans ... everyone. Our home places are unique. We must preserve that uniqueness or we will be poorer, emptier; less whole.

Canada without caribou, or without grizzlies, or without killer whales ... is it still Canada? Or has it ceased to be the real Canada, and grown closer to the emptiness that waits at the end of the last compromise?

If Canada is to have prairie dogs, there must be wild prairies. If the north is to have caribou, there must be old-growth forests. If the Pacific coast is to have rich sea life, there must be clean seas. If we are to live in a future that includes diverse, healthy wildlife populations, then we must act now to insure that their habitat survives into that future.

If you love wild things, it is time to act now. The future will be only as rich as you decide to make it, and only as poor as your inaction allows it to become. If wild animals have enriched your life, what are you going to do to repay them, and to ensure that they continue to enrich the lives of those who come after us?

Selected Bibliography

Banfield, A.W.F., *The Mammals of Canada*. University of Toronto Press, 1974

Burt, William H., and Richard P. Grossenheider, *A Field Guide to the Mammals*. Houghton Mifflin, 1976

Chapman, Joseph A. and George A. Feldhamer, eds., *Wild Mammals of North America: Biology, Management, Economics*. Johns Hopkins Univeristy Press, 1982

Dekker, Dick, *Wild Hunters*. Canadian Wolf Defenders, 1985

Gadd, Ben, *Handbook of the Canadian Rockies*. Corax Press, 1986

Hoyt, Erich, *The Whales of Canada*. Camden House, 1984

Hummel, Monte, *Endangered Spaces*. Key Porter Books, 1989

Leopold, Aldo [1949], *A Sand County Almanac*. Oxford University Press, 1966

Further Information

Some organizations that work to conserve wild animals and the ecological health of their habitats:

Canada

Canadian Nature Federation
   453 Sussex Drive
   Ottawa ON K1N 6Z4
Canadian Parks and Wilderness
   Society
   Suite 1150, 160 Bloor Street E
   Toronto ON M4W 1B9
Canadian Wildlife Federation
   1673 Carling Avenue
   Ottawa ON K2A 3Z1
Sierra Club of Western Canada
   BC Chapter
   Suite 314, 620 View Street
   Victoria BC V8W 1J6
   Alberta Chapter
   Suite 248, 2116 – 27th Ave NE
   Calgary AB T2E 2A6
   Manitoba Chapter
   Goodlands MB R0M 0R0
World Wildlife Fund, Canada
   90 Eglinton Avenue E
   Toronto ON M4P 2Z7

Alberta

Alberta Wilderness Association
   Box 6398 Station D
   Calgary AB T2P 2E1
Federation of Alberta Naturalists
   Box 1472
   Edmonton AB T5J 2N5

British Columbia

BC Wildlife Federation
   Suite 102, 6070 – 200th Street
   Langley BC V3A 1N4
Federation of BC Naturalists
   321 West Broadway
   Vancouver BC V6H 4A9
Western Canada Wilderness Committee
   20 Water Street
   Vancouver BC V6B 1A4

Manitoba

Manitoba Naturalists' Society
   Suite 302, 128th Avenue
   Winnipeg MB R3B 0N8
Manitoba Wildlife Federation
   1770 Notre Dame Avenue
   Winnipeg MB R3E 3K2

Saskatchewan

Saskatchewan Natural History Society
   Box 4348
   Regina SK S4P 3W6
Saskatchewan Wildlife Federation
   Box 788
   Moose Jaw SK R3E 3K2

Metric/Imperial Conversion Table

1 centimetre (cm)	0.394 inches
1 metre (m)	3.28 feet
1 kilometre (km)	0.62 miles
1 hectare (ha)	2.47 acres
1 square kilometre (sq km)	0.386 square miles
1 kilogram (kg)	2.205 pounds
1 tonne	0.9842 UK tons
	1.102 US tons
1 litre (L)	0.22 Imp. gallons
	0.264 US gallons
1° Celsius (C)	1.8° Fahrenheit

0° Celsius = 32° Fahrenheit (F)

About the Author

Kevin Van Tighem was born in Calgary and has been interested in wildlife all his life. He studied plant biology at the University of Calgary and graduated with a B.Sc. with distinction. During eight years with the Canadian Wildlife Service, Kevin was involved in inventory and habitat studies in western Canada's national parks, including Elk Island, Jasper, Banff, Glacier, and Mt. Revelstoke. He has worked for the Canadian Parks Service in Jasper, Yoho and Waterton Lakes National Parks and with the Calgary office.

He lives with his wife and three children in Waterton and enjoys angling, hunting, and outdoor writing. He has published numerous magazine articles on wildlife and conservation.